Tucholsky Wagner Zola Scott Sydow Freud Schlegel
Turgenev Wallace Fonatne
Twain Walther von der Vogelweide Fouqué Friedrich II. von Preußen
Weber Freiligrath
Kant Ernst Frey
Fechner Fichte Weiße Rose von Fallersleben Richthofen Frommel
Engels Fielding Hölderlin
Fehrs Faber Flaubert Eichendorff Tacitus Dumas
Eliasberg Ebner Eschenbach
Feuerbach Maximilian I. von Habsburg Fock Eliot Zweig
Ewald Vergil
Goethe London
Mendelssohn Balzac Shakespeare Elisabeth von Österreich
Lichtenberg Dostojewski Ganghofer
Trackl Stevenson Rathenau Doyle Gjellerup
Mommsen Tolstoi Hambruch
Thoma Lenz Droste-Hülshoff
Dach von Arnim Hanrieder
Verne Hägele Hauff Humboldt
Karrillon Reuter Rousseau Hagen Hauptmann Gautier
Garschin Baudelaire
Damaschke Defoe Hebbel
Descartes Hegel Kussmaul Herder
Wolfram von Eschenbach Dickens Schopenhauer Rilke George
Bronner Darwin Melville Grimm Jerome Bebel
Campe Horváth Aristoteles Voltaire Federer Proust
Bismarck Vigny Barlach Heine Herodot
Gengenbach
Storm Casanova Tersteegen Gilm Grillparzer Georgy
Chamberlain Lessing Langbein Gryphius
Brentano Lafontaine
Strachwitz Claudius Schiller Kralik Iffland Sokrates
Katharina II. von Rußland Bellamy Schilling
Gerstäcker Raabe Gibbon Tschechow
Löns Hesse Hoffmann Gogol Wilde Gleim Vulpius
Luther Heym Hofmannsthal Klee Hölty Morgenstern Goedicke
Roth Heyse Klopstock Kleist
Luxemburg Puschkin Homer Mörike
Machiavelli La Roche Horaz Musil
Navarra Aurel Musset Kierkegaard Kraft Kraus
Lamprecht Kind Moltke
Nestroy Marie de France Kirchhoff Hugo
Nietzsche Nansen Laotse Ipsen Liebknecht
Marx Lassalle Ringelnatz
von Ossietzky May Gorki Klett Leibniz
vom Stein Lawrence Irving
Petalozzi
Platon Pückler Knigge
Sachs Poe Michelangelo Kock Kafka
Liebermann Korolenko
de Sade Praetorius Mistral Zetkin

The publishing house tradition has created the series **TREDITION CLASSICS**. It contains classical literature works from over two thousand years. Most of these titles have been out of print and off the bookstore shelves for decades.

The book series is intended to preserve the cultural legacy and to promote the timeless works of classical literature. As a reader of a **TREDITION CLASSICS** book, the reader supports the mission to save many of the amazing works of world literature from oblivion.

The symbol of **TREDITION CLASSICS** is Johannes Gutenberg (1400 – 1468), the inventor of movable type printing.

With the series, tradition intends to make thousands of international literature classics available in printed format again – worldwide.

All books are available at book retailers worldwide in paperback and in hardcover. For more information please visit: www.tredition.com

tradition was established in 2006 by Sandra Latusseck and Soenke Schulz. Based in Hamburg, Germany, tradition offers publishing solutions to authors and publishing houses, combined with worldwide distribution of printed and digital book content. tradition is uniquely positioned to enable authors and publishing houses to create books on their own terms and without conventional manufacturing risks.

For more information please visit: www.tredition.com

The Wonders of the Jungle, Book Two

Sarath Kumar Ghosh

Imprint

This book is part of the TREDITION CLASSICS series.

Author: Sarath Kumar Ghosh
Cover design: toepferschumann, Berlin (Germany)

Publisher: tradition GmbH, Hamburg (Germany)
ISBN: 978-3-8491-5061-7

www.tredition.com
www.tredition.de

Copyright:
The content of this book is sourced from the public domain.

The intention of the TREDITION CLASSICS series is to make world literature in the public domain available in printed format. Literary enthusiasts and organizations worldwide have scanned and digitally edited the original texts. tredition has subsequently formatted and redesigned the content into a modern reading layout. Therefore, we cannot guarantee the exact reproduction of the original format of a particular historic edition. Please also note that no modifications have been made to the spelling, therefore it may differ from the orthography used today.

TO THE CHILDREN

My dear, I am now going to tell you many more Wonders of the Jungle, as I promised to do in Book I.

In that Book, as you will remember, I promised to tell you more about the elephants and about the laws of their herd. So I shall do so now.

Then I shall tell you about some animals which I did not describe in Book I. Among these you may like to know especially about the tiger, the lion, the leopard, and the wolf.

You may like to know how really *clever* some of these animals are, and how some of them have *affections*, just as we have.

But while you are reading about them, you must try to *think*. Then you will understand *why* these animals do certain things. And that will show how clever *you* are!

I have used a few new words in this Book. But I am sure you know them already.

Now I shall begin with the laws of the elephants.

[v]

CONTENTS

I. The Elephant Herd a Republic
   The Duties of the President
   He Must Provide Daily Food
   He Must Provide Daily Drink
   He Must Keep Order in the Herd
   He Must Avoid Danger from Outside

II. War and Neutrality in the Jungle
   Wise Elephant Leader Avoids War
   Wise Elephant Leader Keeps Neutral
   When it is Impossible to Remain Neutral

III. The Policemen of the Elephant Herd

IV. The Punishment of the Wicked Elephant
   The Princes and the Bad Elephant
   The Trial of the Criminal Elephant—as in a Court of Law
   The Infliction of the Punishment
   The Rogue Elephant
   The Brand of the Rogue
   The Reward of Repentance

V. Flesh-eating Animals: The Felines, or the Cat Tribe
   The Feline has Strong Fangs
   The Feline's Tongue is Rough
   The Feline's Claws are Retractile
   The Feline has Padded Paws

VI. The Tiger [vi]

The Life History of the Tiger Family
The Tiger's Family Dinner

VII. The Tiger Cubs' Lessons
Tiger Cubs Learn to Kill Prey, After their Parents have Caught It
Tiger Cubs Take Part in Hunt to Catch Prey
Tiger Cubs Learn to Catch Prey by Themselves

VIII. The Tigress Mother's Special Duties
The Truce of the Water Hole

IX. The Special Qualities of Tiger and Tigress
Both Tiger and Tigress Defend Their Cubs
The Tiger Family's Lost Dinner
The Tiger as a Heroic Husband

X. The Lion
The Lion Has the Fangs, the Tongue, the Claws, and the Paws of a Cat
How the Lion is Different from Other Cats

XI. The Lion's Daily Life

XII. The Lion a Noble Animal
Androcles and the Lion
The Lady and the Lioness

XIII. The Leopard
The Leopard's Ground Color and Spots
Why the Leopard has Spots

XIV. The Leopard's Habits
   The Panther: Popular Name For Large Leopard
   How the Leopard Seizes his Prey [vii]
   The Leopard's One Amiable Quality—He Loves Perfumes
   The Leopard and the Lavender

XV. American Leopard: The Jaguar

XVI. The Dog Tribe
   The American Gray Wolf
   The American Wolf Learns to Evade the Gun
   The American Wolf Learns to Evade the Trap
   The American Wolf Learns to Evade the Poison

[viii]

ILLUSTRATIONS

Elephants at Work

Elephant Leading Herd Through the Jungle
Trained Elephants at the Court of a King
Elephants Guarding a Bad Elephant
Policemen Elephants Arresting a Criminal Elephant
Good Elephant Heading off a Criminal Elephant
Tiger
Tiger Protecting his Cub
Tiger Charging Hunting Party
Group of Lions
Puma
African Lion
Giraffes
Kangaroo
Androcles and the Lion
Leopard
Jaguar
The Chain of Conflict in the Jungle
Gray Wolf

[1]

THE WONDERS OF THE JUNGLE

CHAPTER I

The Elephant Herd a Republic

An elephant herd is a kind of republic, something like the United States of America, only much smaller and much simpler. So its leader is a sort of president. He is usually the wisest elephant in the herd.

You may like to know how the elephants choose their president. I shall tell you how they do that.

But you must first consider how the people of the United States choose *their* President. They find out who among their important men is best able to lead them in all the great duties of the nation. Then they choose *him*.

But if afterward they find that he is *not* leading the nation in the wisest manner, then the people of the United States choose another man to be their President the next time. [2]

The elephants in a herd do something like that. They first follow the elephant who, they think, is best able to lead them. But if afterward they find that he is not leading them through the jungle in the right way, and that another elephant could lead them in a better manner, then they follow him instead. He then becomes the president of the herd.

"But what is the best way of leading the herd through the jungle?" you may ask.

I shall now tell you about that. The best way to lead the herd is *to satisfy all their needs*. So the president of the herd has four great duties.

The Duties of the President

First Duty: He must lead the herd in such a manner that all the elephants will get enough *food to eat* every day.

Second Duty: He must lead the herd in such a manner that all the elephants will get enough *water to drink* every day.

Third Duty: He must *keep order* in the herd, and not allow any naughty elephant to fight or quarrel.

Fourth Duty: He must guide the elephants in such a manner as to *avoid all danger from outside*; [3] and if such danger does happen to come, he must guard the herd from that danger.

I shall now tell you about these four duties more fully.

He Must Provide Daily Food

Elephants are such large animals that they need a great amount of food. So they have to walk a long way every day, munching the leaves of the trees as they go.

They walk in line, one behind another, as that is the easiest method of walking through the thick jungle; for then one gap through the jungle is enough for all the elephants to go through, one at a time, and they need not make a different gap for each elephant.

Now you will understand that if that one gap is big enough for the *largest* elephant to go through, it is of course big enough for *all* the elephants to go through. So, if the largest elephant walks first, in front of the line of elephants, he can force a way through the thick jungle that will be big enough for all the other elephants who come behind him.

So usually the largest and strongest bull elephant is the leader of the herd—if he also [4] has the other qualities of a president, which I shall presently describe more fully. To have all the qualities of a president, he must not only be strong, but also wise and clever. Why? Because even in merely going through the jungle a wise leader avoids many difficulties. It might be that the jungle straight ahead was very thick, and it would be hard to force a way through it; but by turning a little to the right or to the left, an easier passage could be made. This a wise leader would find out, and then turn in that direction.

Again, in the jungle, the ground is sometimes too soft; it might be made of clay which had become soft owing to rain a few days before. But elephants are such heavy animals that they cannot go far over soft ground, as their feet would sink in too deep. And the ground might be covered with bushes or tall grass, so that the elephants could not *see* to what distance the ground was soft. They might not mind going over soft ground for a few yards, but they would not like to go over such ground for a whole mile.

[5]
[6]

Elephant Leading Herd through the Jungle

So a wise leader would know by glancing around how far the ground was likely to be soft; and if he learned that it was likely to be

[7] soft for a large area, he would turn at once and go around it. But a foolish leader might take the herd right into the soft ground, and they would all be stuck in the mud, and have a lot of trouble getting out of it again.

So if the herd has chosen merely the biggest and strongest elephant to be their president and he makes such mistakes as that, they soon depose him; that is, they no longer follow him. They look around for some other leader who can discover a better way, and they follow him instead. And if afterward they find that he is wise and clever, and does not make mistakes, they follow him as their leader every day after that, even if he is not quite so big and strong as the other elephant was.

He then becomes the new president, if he is at least strong enough to make a good gap through the jungle. Most of the elephants could pass through that; only the biggest bull, the deposed president, would have the trouble of enlarging the gap with his body in going through it. And this would serve him right!

In the same manner the leader of the herd must not go over ground that is *too hard*, for elephants are such heavy animals that it jars the bones of their feet to go over hard ground [8] for a great distance. If there has been no rain for several weeks, then in a hot country the ground gets very hard in some places. So if there has been no rain near a herd for some time, a wise leader avoids these hard places.

So, as you see, an elephant leader has to be quite clever in merely avoiding difficulties, in the daily search for food. And that is not all! The food itself may be plentiful in one part of the jungle, and rather scarce in another; for in one direction there may have been just enough showers recently to bring out the fresh leaves on the trees; but in another direction there may have been no rain at all for some time, and so there would be no fresh leaves there.

Why, even in your own town there may be a good shower of rain in one part of the town, and no rain at all in another part. So it might be in the jungle; a wise leader would know this by instinct, and he would take the herd along that part of the jungle where there had been recent showers of rain, and where there would be enough fresh leaves. [9]

He Must Provide Daily Drink

After the elephants have had enough to eat for the day, they must have enough clear water to drink. And to get this is *the hardest daily duty of the leader*.

In the jungle, even if the leader makes a little mistake and goes the wrong way, there may still be enough to eat, because the elephants can always find enough trees in the end by going a little farther: so they would have only a little more trouble in getting their food, if the leader made a mistake. But with *water* it is quite different—the leader may find no water at all, if he makes a mistake and leads the herd the wrong way.

"Then how must he lead the herd so as to find water, as well as food?" you may ask.

I shall tell you. In most jungles there is a river or even a small stream from which the elephants can drink. But the river or stream may go winding in and out of the jungle, so that it is in one part of the jungle but not in another part. So a wise leader tries to keep his herd near one of those parts of the jungle through which the river flows.

In fact, if the elephants and even the other [10] wild animals are lucky enough to find a fairly big river, and the jungle near that river has plenty of food in it, then the animals stay near there almost all the time. They eat from the jungle and drink from the river; and sometimes they come to the very same place to drink—as at the Midnight Pool, which I described to you in Book I.

So if the leader of the elephant herd is lucky enough to find such a jungle, with plenty of food and a big river in it, he keeps the herd there all the time; and then they have no more trouble about food or drink.

But suppose the leader cannot find such a place? Suppose there is a river, but not enough food near the river? Then what does a wise leader do?

He leads the herd in such a way as to make *a kind of curve*. He goes into the jungle by the easiest way in the beginning; then, after the elephants have eaten a little, he starts turning slightly toward the direction in which the river flows. When the elephants have eaten a little more, he turns still more in that direction.

In this manner he leads the herd in a kind of curve toward the river, browsing all the way from the trees near by. So, at the end of the [11] day, when the elephants have had enough to eat, they reach the river and have also enough to drink. Is not that a very clever method of providing both food and drink for the herd?

If the herd sleep near the bank that night, they start from there the next morning in their search for food; and they usually go into the jungle by the same path by which they came. But on *returning* to the river to drink that night, the leader need not bring them back by exactly the same path.

The fact that they did not have enough to eat right near the river shows that the jungle is not very thick there; so the elephants will have no trouble in making a fresh path, a little higher up the river, or a little lower down. A wise leader usually does that: he leads the herd to the river slightly higher up or lower down, and so he makes a slightly different curve through the jungle. Why? Because if he kept to exactly the same curve from the jungle to the river every day, the herd would eat up all the leaves along that path in a few days. So, by changing the curve a little from time to time, he allows fresh leaves to grow there meanwhile.

You now understand why the president of [12] the elephant herd must be wise and clever to do all that I have told you so far. Even among men the President of a Republic has similar duties to attend to, though in a different manner: he too has to govern his country in such a manner as to provide the people with their daily wants, if they obey the laws and do honest labor.

In the elephant herd everyone has to do honest work, as he has to gather his own food; and he has also to obey the laws of the herd. I shall now tell you about that.

He Must Keep Order in the Herd

The third duty of the elephant leader is to keep order in the herd. Most elephants are by nature gentle, docile, and obedient. That is why men can tame them and make them work; otherwise, if elephants were by nature fierce and disobedient, men could not train them so perfectly as to perform at a circus, or carry people in a procession. So even in the jungle, where the elephants are wild, they usually obey the leader and keep the laws of the herd.

These laws chiefly concern their daily food and drink. As I have told you, in their daily search for food the elephants march in a line, [13] one behind another. A selfish elephant in the middle of the line might want to stop and eat up *all* the leaves on a tree near him; and if he did so, he would block the way for those behind him, and besides, there would be no leaves on that tree for them to eat when they came to it.

So there is a general rule in the herd that each elephant must take just a few of the leaves from a tree, and then *move on*; and if instead he does block the way, the elephants behind him may push him forward and make him move on.

"But," you may ask, "why can't the other elephants behind him also stop and eat up all the leaves on the trees near them?"

Because then all the trees on that line of march would be bare of leaves, and it might take a whole month for fresh leaves to grow there again. But if the herd took only a portion of the leaves from each tree, there would be enough food for them along that path if they happened to visit it again in a few days.

In fact, the elephants need make only a few such paths through the jungle, if they eat only part of the leaves at a time along any of the [14] paths. Then they can visit these paths in turn on other days, and always find enough food there—because the fresh leaves constantly growing on the trees would make up for the small portion they had eaten.

So you understand how wise the elephants are in having that law in the herd.

"But," you may say, "if they were to eat *all* the leaves on a tree, their path would be a short one; while if they eat only a portion of the leaves, their path would be much longer, as they must nibble from many more trees to satisfy their hunger."

That is quite true. But there is no advantage in having a short path, because at the end of their march in search of food they must find water to drink, as I have already told you—and they may have to go several miles to reach the nearest stream. So they might as well nibble from the trees all the way to the stream, especially as elephants can easily march ten or twelve miles in that manner every day.

Besides, after taking a bunch of leaves from a tree, they must chew it before taking the next bite; so, meanwhile, they might just as well walk on to the next tree. In fact, if they [15] have not quite finished chewing, most elephants pass by one or two trees before taking the next bite. That shows how really wise they are. For then they are *sure* of finding enough food along that path when they visit it again a few days later.

It is the president of the herd who sets a good example to the others in doing all these wise things. As he walks at the head of the line, he sees at a glance what is the best thing to do in that particular path, whether to nibble a little from every tree, or to pass by a few trees without nibbling from them at all. And whatever he does, all the other elephants do after him.

My dear children, it is exactly the same among us. When food is scarce in a country and people must be careful, then it is the President who tells us how to portion out the food supply in the country. Otherwise, some people would be wasteful and throw food away—and others would not have enough to eat.

It is very important to learn from your childhood to be careful of food. Do you know that in the United States every man, woman, and child on an average throws away every year seven dollars' worth of food *on the plate*? [16] That would be enough to feed all the people in the poorhouses and the hospitals.

Elephants are most careful of their food. Their president is all the time thinking of the best method of making the food supply of the jungle last them from season to season. But the other elephants must help him to do that, by following his good example. If any particular elephant is selfish and wants to eat up at once all the food near him, he is pushed out of the line by the other elephants, as I have already told you. If he is naughty again, he is more severely punished.

How he is punished, I shall tell you in another chapter. I shall then tell you how all sorts of naughty elephants are punished; for, just like people in a country, I am sorry to say that there are in the jungle a few elephants that do not obey the law.

An elephant can be selfish not only in eating, but also in drinking. You will remember what I told you in Book I—how all the elephants stand in a line along the bank of a stream and drink; and after they have all satisfied their thirst, they may jump into the water to bathe and swim.

It would be very selfish for an elephant to [17] jump into the water before the others had finished drinking; for then he would muddy the water which some of the others were still drinking. And for such conduct an elephant is very severely punished.

But the very worst offense in an elephant herd is quarrelling and fighting; for, sometimes, two elephants do quarrel and fight, just like a couple of naughty boys in school. But there is never any real need to quarrel in an elephant herd; for if one of the elephants has done wrong or broken the rules of the herd, he will be punished by the president of the herd—just as in school a naughty boy would be punished by the teacher or by the head of the school.

It is not necessary for any other elephant in the herd to quarrel or fight with the naughty elephant, even if he has been injured by him; the president of the herd will punish the naughty elephant soon enough. So if two elephants do fight, *both* of them are punished; of course the one who began the fight is punished more severely than the other. [18]

He Must Avoid Danger from Outside

The president of the herd must lead the elephants in such a manner as to avoid any danger that may come to the herd from outside. In the jungle there are other wild animals; most of them are, of course, too small to be able to hurt so large an animal as an elephant; but a tiger is so strong and so fierce that he could kill a small, half-grown elephant.

The tiger could hide in the jungle, and if the small elephant happened to stray from the herd, the tiger could spring upon it and kill it. So the president of the herd usually keeps the elephants away from any part of the jungle which he knows to be infested by tigers.

How does he know that? By the paw marks made on the ground by the tigers. For the tigers leave plenty of paw marks on the ground in coming in and out of their dens to hunt their prey every day. So if the president of the elephant herd comes across a line of such paw marks, he turns aside and leads the herd in another direction.

Of course, if the herd happened to meet a tiger quite suddenly, they would at once face the tiger. And the tiger would never dare to [19] attack even the smallest elephant if the big ones were near, for they could drive him off with their tusks or trample upon him.

But the greatest danger that can come to an elephant herd from outside is from men. Men sometimes go into the jungle to shoot wild elephants with guns, or to catch them alive in huge traps. So the leader of the herd must find out where the traps are, or where the hunters are hiding; and then he must avoid such places.

You will remember what I told you about Salar and his father in Book I. Salar was the boy elephant who nearly fell into a most tricky trap, but his wise old father suspected the trap and called to Salar to halt; and because Salar obeyed his father and halted at once, he just escaped falling into that awful trap.

Well, in the jungle hunters lay all kinds of traps to catch wild elephants alive; and sometimes for several years the hunters try over and over again to catch the elephants, if they fail to catch them at once. So the president of an elephant herd has to look out for traps all the time; and the herd that has the wisest president escapes capture for the longest time. [20]

In fact, as Salar is an actual elephant, not an imaginary one, I may tell you that his father was such a wily leader of his herd that he kept them from capture for ten years longer than the leader of any other elephant herd in that jungle.

As for hunters who seek to kill wild elephants with guns, the leader of the herd has to be even more careful in avoiding them.

These hunters usually hide behind bushes, and try to creep up to the elephants; and when they are within a hundred yards of the elephants, they begin shooting them. Then the leader of the herd has to prove his wisdom.

A foolish leader would stand still, or even try to charge the hunters; and then more of the elephants would get killed. But a wise leader gives the signal to *run away* as soon as he hears the sound of the first gun; then at most only one or two of the elephants are killed—and sometimes none at all.

Why? Because to kill an elephant with a gun a hunter must hit him exactly in one particular place on the body—behind the elephant's ear, where the skin is thin. At the first shot the hunter may not hit the elephant just there, but inflict only a trifling wound [21] elsewhere on his thick skin. So by running away at once an elephant may save his life.

But as all leaders are not so wise, the hunters usually manage to kill one or two of the elephants. I may tell you that these hunters kill the elephants merely to get their tusks, which they sell as ivory.

It is a shame to kill such wonderful animals just for money; and you ought to know that in some parts of Africa almost all the elephants have now been killed. If the hunters continue to do that, there will be no elephants left in Africa in a few years. Then the hunters will not be able to get the very ivory for the sake of which they now kill the elephants.

But you will be pleased to know that in India and other countries of Asia nobody is allowed to kill a wild elephant; for if anyone did so, he would be put into jail. Special hunters are allowed to catch wild elephants alive, as I have already told you; and then the elephants are tamed and trained to do all kinds of useful work, such as to pile logs, build bridges, make roads, and lay water-pipes (see Frontispiece). Some of these elephants are also taught to do tricks in a circus, or to carry grand people in a procession. [22]

"Then how do people in India get their ivory, if they never kill an elephant?" you may ask.

They get the ivory when the elephant dies naturally; and the ivory is just as good then as before. Is not that very wise? The people of

India first get the help of the elephants in doing all their heavy work, and at last they get the ivory also.

There are huge buildings in India, some of which are more than two thousand years old, which are so wonderful that engineers in America and Europe do not know exactly how those buildings were erected. There is a particular temple on the top of a mountain; and that mountain is 6000 feet high. The ceiling over the center of the temple is a huge circular piece of marble; and that marble ceiling is so large that for a long time people in America and Europe did not know how it was dragged up to the top of the mountain, and then placed over the temple. But now we know that a team of trained elephants was used to do that.

[23]
[24]

Trained Elephants at the Court of a King

You will be pleased to know, too, that the people who built that temple are called Jains, whom I mentioned in Book I, page 163 (footnote), as the people who are kind to all animals, [25] and who never hurt even the smallest insect. Instead, these mild and gentle people have taught dumb animals to help them build one of the greatest wonders of the world.

How the elephants were taught to do that, I shall tell you in the next Book.

[26]

CHAPTER II

War and Neutrality in the Jungle

Now I must tell you about another duty of the president of the elephant herd: he must avoid another kind of danger that may come to the herd from outside.

I am sorry to say that herds of elephants sometimes fight with one another, just as nations of people do. Alas, although elephants are usually such wise animals, they are sometimes as foolish as men! Two herds of elephants may find the same feeding ground, which has plenty of trees to eat from, and a convenient stream of water to drink from. Then the two herds may start fighting for that new feeding ground—just as two nations sometimes fight for a new land.

Among elephants the herd that first finds the feeding ground usually keeps it; but another herd may come there at about the same time, and claim to have found it first—and may fight the other herd for that new feeding ground. Or it may happen that the second herd really [27] came there later, but is stronger than the first herd, as it has more bull elephants in it. Then the second herd may try to drive away the other herd, which really found that feeding ground first.

Wise Elephant Leader Avoids War

Then what does the president of the first herd do? Alas, he usually stays there to fight it out. But he gains nothing by it; instead, some of his bulls get killed or wounded—and in the end his herd has to flee just the same. A very wise leader would have done that from the first; for he might find another feeding ground just as good somewhere near. And besides, the quarrelsome herd will be punished soon enough!

"How will it be punished?" you may ask.

I shall tell you. A quarrelsome herd gets into the *habit* of quarrelling with other herds, just as a quarrelsome boy gets into the habit of quarrelling with everybody—or even as a quarrelsome military nation gets into the habit of quarrelling with other nations. Then that quarrelsome boy might meet a stronger boy some day — and get a good thrashing! And the quarrelsome nation might attack a more powerful nation some day — and get a good thrashing!

So also that quarrelsome herd of elephants might some day attack a herd which proves to be stronger. Then that naughty herd would also get a good thrashing. So it is foolish, indeed, for the president of a herd to domineer over weaker herds in the jungle.

Indeed, there is a still greater punishment for a quarrelsome herd. I have already told you that there are hunters who lay traps to catch wild elephants alive. Well, these hunters try specially to catch a quarrelsome herd first! Why? Because quarrelsome herds kill or injure other wild elephants with whom they fight. But the hunters do not want to have any of the elephants killed or injured, as they want to catch as many of them as possible in order to teach them to do useful work. So they catch the quarrelsome herd first, before it can kill or injure many of the other elephants.

Of course, the hunters know which is a quarrelsome herd, because they send men into the jungle from time to time to watch different herds and keep track of them.

The Wise Elephant Leader Keeps Neutral

There is still another duty that the leader of the elephant herd must do. Sometimes it happens that as he is taking his herd through the jungle, he meets two other herds that are fighting. Then what must he do?

He must lead his herd by another path. He must not take part in the fighting between the two other herds. He must keep *neutral*.

What does that mean? It means that he must not meddle with other peoples' fights and quarrels. He must not take sides; that is, he must not help either of the herds to beat the other. That is the usual rule in the jungle which a wise elephant leader tries to keep.

But there is an exception to that rule. It sometimes happens that it is impossible for the president of an elephant herd *not* to take sides. When does that happen? I shall tell you.

When it is Impossible to Remain Neutral

When two herds are fighting, they may get very reckless. When men make war, they knock down houses with their guns, and trample on growing corn. In the same manner, when two herds of elephants fight they knock [30] down trees, and trample on shrubs and bushes — sometimes the very trees and shrubs and bushes for which they are fighting! *There never is a fight of any kind without a lot of damage being done.*

So it may happen that one of the fighting herds gets so reckless that it comes into the ground of the herd that has kept neutral, and does a lot of damage there. Then what must the president of the neutral herd do? He must defend his own ground from damage.

So long as the fighting herds kept to their grounds, he must not interfere. But when one of the fighting herds comes into *his* ground and does damage, he must defend his rights. A wise elephant leader always does that; for he has bull elephants of his own who can drive out the intruders.

[31]

CHAPTER III

The Policemen of the Elephant Herd

I have already told you that the president of an elephant herd must keep order within his own herd; that is, he must not allow a naughty elephant to commit a crime, such as to attack any other member of the herd. And if a naughty elephant does commit a crime, it is the duty of the president to punish him.

I shall now tell you how he does these things. *There is a wonderful police system in an elephant herd.*

You will understand that better if I tell you first about an old police system among men. You will read in history books about the Anglo-Saxons, who were the forefathers of most of the people of England and of the United States of to-day. These Anglo-Saxons had a police system like this:—

In a village or in a town all the grown-up men were divided into groups of ten men; and if any man tried to commit a crime, all the other nine men of his group tried to prevent him. If he committed the crime *before* the other [32] nine men could prevent him, they at least arrested him. Then they took him before the judge for punishment.

It is something like that in an elephant herd in the jungle; only, as there are not so many bull elephants in a herd as there are men in a village, it is not necessary to divide the bulls into different groups.

As there are only twenty or thirty grown-up bulls in an average elephant herd, it is the duty of *all* the grown-up bulls to prevent a bad elephant among them from committing a crime; and usually it is the bulls nearest to him who actually stop him from committing the crime. If he manages to commit the crime *before* they can prevent him, they surround him immediately and keep him there like a prisoner, till the president of the herd comes to punish him.

My dear children, that is a great lesson for us. A good citizen always helps to keep the law; if he sees anyone breaking the law, he tries to prevent him from doing so. Some men do nothing, if they see a person breaking the law; they say, "It is no business of ours."

Elephants are much better citizens of the jungle in that respect; they always try to prevent a bad elephant from breaking the law.

[33]
[34]

Elephants Guarding a Bad Elephant

[35] Now I am going to tell you something that will astonish you—as it has astonished a good many clever scientific men. Do you know why people are at all able to use elephants in a circus, and give you pleasure by making them do tricks? Suppose one of the elephants suddenly went mad? Then he could kill a dozen people in a minute by just rushing at them and trampling on them. No *men* could stop him, even if they had guns ready all the time; for it might take several minutes to kill an elephant even with a special kind of gun. And meanwhile the mad elephant could trample upon scores of people in a crowded circus.

And it is just the same in a procession, when elephants are used to carry grand people—kings and queens, princes and princesses, lords and ladies. An elephant in a sudden fit of rage could kill many of them.

Then why do people use elephants in a circus or in a procession? Why do they trust themselves with such large and strong animals? Just think!

"Because an elephant is naturally docile and gentle," you may say.

That is quite true. But still a bull elephant might get into a *sudden* fit of rage about some [36] thing, just like a naughty boy; and as a naughty boy in a sudden fit of rage might break things, so also that bull elephant might rush about and trample on people.

Then why do people trust themselves with elephants? Think again!

It is because of the *police system among the elephants themselves*. Because if any elephant in a circus or a procession tried to do any mischief, even in a sudden fit of temper, all the other elephants there would prevent him! The *men* there might not be able to prevent him; but the other *elephants* could, and they would.

Nobody need tell the other elephants to do that. Without being told to do so, they would rush to him, surround him, and prevent him from doing any mischief. And if only one bull elephant happened to be near enough to him at that time, he would at least head him off—that is, throw himself in the way of the angry elephant. I shall tell you a wonderful story about that presently.

I have said that nobody need tell the other elephants to prevent a bad elephant from committing a crime. The other elephants would do that themselves, because *they have got into the habit of doing so in the jungle*. [37]

I must tell you that almost all the elephants you see in a zoo or a circus were once wild in the jungle; they have been caught, then tamed, then trained. But they still remember the laws of the jungle; and they follow those laws whenever necessary—just as children who get into the habit of keeping the rules of their school also form the habit of keeping the law when they grow up. So the men who use elephants allow them to practice this particular law; that is, they allow and encourage the elephants to continue this police system among themselves.

From this you will understand that people do not usually use a bull elephant singly; that is, they usually use a number of bull elephants together, so that all the others would prevent a bad elephant from doing any sudden mischief.

Wise people who know the habits of elephants usually use a number of them at a time. But there have been many foolish people who have used a bull elephant by himself; then somebody has ill-treated that elephant, and in his rage he has done a lot of harm.

That actually happened in a big zoo recently. Then they had to shoot the elephant. That shows that the people at that zoo knew very [38] little about the habits of elephants. They should have kept that elephant with a few other elephants.

You may like to know how wise people in Europe and America have learned the habits of elephants. They learned them from the people of India many centuries ago. The people of India first observed wild elephants in the jungle; and they discovered that the elephants had wonderful laws in their herds—which I have described to you. Then the people of India caught the wild elephants, and tamed them, then trained them to do tricks and also useful work.

About 2250 years ago there was a famous king in Europe named Alexander, who went to India. There he and his followers saw the wonderful things that the people of India had taught the elephants to do. So they brought some of these people to Europe, with their elephants. That is how the people of Europe first learned about the wonderful habits of elephants. In our own times, wise people who bring elephants to Europe and America also bring a few men who know the habits of elephants.

That is why it is such fun to watch the elephants at a circus. [39]

CHAPTER IV

The Punishment of the Wicked Elephant

Now I shall tell you how naughty elephants are punished. I have already told you that if a naughty elephant attacks any other elephant in the herd, all the other bulls surround him and keep him there, till the president of the herd comes and punishes him. Now I shall tell you how that is done.

The bull elephants stand in a ring a few yards away from the culprit; but they all face him, so that they can watch him all the time. Then the president of the herd steps into the ring, and walks toward the back of the culprit.

"But if the culprit keeps turning round, so that the president cannot get behind him?" you may ask.

Then two of the bulls forming the ring step in; and they come and dig the culprit in the ribs with their tusks, one on the right side and the other on the left side. Then the culprit cannot turn; he must stand still and take his punishment.

And this is the way the punishment is given. [40] The president gores him with his tusks on the hind quarter, just as a father spanks his naughty boy—only much harder! In fact, after two or three blows from the president's tusks, the culprit's back is very sore.

How long does this punishment last? Well, just about as long as the spanking of a naughty boy by his father. How long is that?

"Till he says he is sorry, and won't be naughty again," you may say.

That is exactly what happens to the bad elephant. The president goes on goring him till he *says* and *shows* that he won't be wicked any more. Yes, an elephant can *say* that he won't be wicked again by whining; and he can *show* it by the way he holds his head and trunk. You will understand that better from the story I shall now tell you. It is a true story. It is about a bad elephant in the service of men after the elephant had been tamed; but the punishment for being wicked would have been just the same if he had been a wild elephant in the jungle.

The Princes and the Bad Elephant

It happened a few years ago, when King George and Queen Mary of England went to [41] India. At that time a young reigning prince in India had just succeeded to his father's throne. So there were many ceremonies at the palace, and festivities among the people. These functions lasted a whole week, and several elephants were used in processions.

One day the elephants were taken to a place ten miles away to do useful work, such as to pile timber for building a bridge. Among these elephants was one called Mukna.

Mukna was a bad-tempered elephant. His tusks never grew more than half-size. Bull elephants whose tusks do not grow to their full size are sometimes bad-tempered; they seem to have a grudge against everybody. Such elephants are always treated with special kindness, as if to make up to them for their loss.

But in spite of all the kindness Mukna received, his temper grew worse and worse. He was punished for that, though very lightly; he was merely deprived of delicacies in his food. Elephants in the service of men usually get hay, grass, and leaves to eat; but on special days they get sugar cane, bananas, and a kind of pancake, all of which are great delicacies to an elephant.

Mukna's keeper had deprived him of these [42] delicacies for his bad temper, just as a naughty boy's father may deprive the boy of ice-cream. That should have been a lesson to Mukna to be good. But it was not. Instead, he got worse.

One morning, when all the elephants were working, Mukna's keeper ordered him to lift a log. Mukna did not obey. He merely stood still.

Now, disobedience is a serious fault in an elephant — just as it is in a child. In fact, it is the beginning of all faults on earth, as the Bible says. If people once allowed even an elephant to be disobedient, they could not control him any more — just as if a naughty boy were to be left unpunished for disobeying his parents or teacher, he would get worse, and disobey his superiors, and even the law, when he grew up.

So Mukna's keeper looked at him sternly and said, "I command you for the second time to lift that log!"

But Mukna would not yet obey. He merely stood still.

Then all the other elephants looked up from their work, just as grown-up men in a workshop look up if they hear the foreman scolding a bad workman. Those other elephants knew what an awful crime disobedience was. [43]

Then in a deep and stern voice Mukna's keeper said to him, "I command you for the third and last time to lift that log!"

But for the third time Mukna refused to obey.

"Then you shall hear about this!" the keeper said, just as if he were talking to a disobedient workman.

The keeper did not say anything more. But two of the nearest bull elephants stepped up to Mukna, one on each side of him—just like a couple of policemen arresting a criminal. Then a third bull came up in front of Mukna, and stood with his back to him, so that all three police elephants faced the same way as Mukna—as you see in the picture on page 45.

Then at the same time the three police elephants stepped *backward*, so that Mukna also was forced to step backward. Step by step the three police elephants went backward till Mukna's hind legs came against the trunk of a tree. There Mukna was held for a moment, so that he could not wriggle away. For the elephant in front prevented him from moving forward, and the tree prevented him from moving backward; and the two elephants on the sides prevented him from moving sideways. [44]

Then the keeper stepped to the tree and fastened one of Mukna's hind legs to the tree with a chain—so that he could not run away. The three police elephants then went back to their work.

Now I must tell you that in a herd in the jungle a bad elephant is punished at once by the president. But it is slightly different among elephants in the service of men, because there they have no elephant president, but a man president, who might be away at that time. That man is called the elephant master.

That is just what happened when Mukna was disobedient. The elephant master happened to have gone to the palace on a visit. So Mukna's keeper called a messenger and sent him to the palace to report Mukna's disobedience. The messenger had to ride on another elephant to go that distance.

Mukna saw that elephant going toward the palace with the messenger. Mukna knew why! It was to fetch the elephant master, who would punish him! Even a dog that has been naughty will cringe and whine at the sight of a whip, because it knows that its punishment is coming.

[45]
[46]

Policemen Elephants Arresting a Criminal Elephant

But Mukna did not cringe and whine. Instead he became defiant—just like a very bad [47] boy. He held up his head and curled his trunk tight in a spiral in front of his chest. In an elephant that is a sign that he is defiant or determined, just like a man who folds his arms tight across his chest. Mukna was unrepentant.

The messenger reached the palace and reported Mukna's disobedience; and the elephant master said that he would come that afternoon to punish Mukna.

The reigning prince said that he also would come. As he had just ascended his throne, he wanted to teach a lesson to all criminals in his domain from the beginning of his reign, and Mukna was the first to commit a crime in the prince's reign. For, I must tell you, all elephants in service in India are treated just like men; they are rewarded as good citizens or punished as criminals. So Mukna was regarded as a criminal.

The prince asked three other young princes, his cousins, to come with him. A young American was then staying in the palace as a guest, and he also was invited to come.

That afternoon the royal party went with the elephant master to the place where the elephants were; there were about thirty bulls, besides Mukna. The place was a clear space, [48] about a hundred yards across, with a lot of trees along the sides. Mukna was tied by the hind leg to one of those trees.

The royal party got out of their carriages and entered the open space on foot, quite near the spot where Mukna was tied up. They were not thinking of Mukna just at that moment, as they were talking of the grand feasts at the palace. So they did not notice Mukna at once.

Meanwhile Mukna had been brooding all day. He knew that his punishment would come very soon. "I will do it—I will do it!" he must have been saying to himself all the time. In that way he had worked himself into a fury.

When the royal party entered the open space, the young American happened to be nearest to Mukna. As he had just arrived from America, he did not know much about elephants; so the young American did not notice that Mukna was chained up to the tree by the hind leg, and that *he* was the bad elephant they had come to punish. Instead, the young American thought that Mukna was just one of the ordinary tame elephants working there.

So as the royal party happened to pass about [49] ten yards in front of Mukna, the young American stepped aside and said, "Hello, I must pat you!" Saying that, he raised his hand and stepped toward Mukna to pat him.

But meanwhile, when Mukna had seen the elephant master arrive with the royal party, he knew that the moment of his punishment had come! "I will do it—I will do it!" he had kept saying before. So when the young American raised his hand, Mukna suddenly made up his mind *to do it now*!

Mukna gave just one short trumpet. The next instant he gave a vicious tug with his hind leg—and snapped the chain! With a huge stride he came toward the American and the royal party. He would "do it" now! *He would kill them all!*

Nothing could stop him from doing it, it seemed. He would knock them down and trample them to death.

But meanwhile the elephant master had heard the trumpet Mukna had given a moment before he broke the chain. And in an instant the elephant master realized what would happen.

"Run for your lives!" he shouted to the young American and the four princes. And he ran himself. [50]

But an elephant can run much faster than any man. It seemed that nothing could save those six men; they would all be trampled to death. The only direction in which they could run was toward the middle of the open space—away from Mukna. Even if they reached it, they would still have to run toward the trees on the far side. Could they reach the trees in time? No! Mukna was gaining upon them. It seemed that in a few more strides Mukna would hurl himself upon them, and there was nobody to stop him.

But yes—there was!

For meanwhile, just as the elephant master had heard the trumpet Mukna had given, all the thirty bull elephants had also heard it. Most of them were too far off, near the line of trees; but there happened to be a bull a little nearer the middle of the open space. He saw at once that he could not overtake Mukna, if he merely chased him. So, how could he stop Mukna from murdering the six men?

I shall tell you. This is what that bull elephant did. As soon as the men had started running, he saw in what direction they were going. So he turned slightly, and ran also *in that direction*. As Mukna gained upon the men, he too came nearer and nearer to the men.

Good Elephant Heading off a Criminal Elephant

Mukna had come within three yards of the young American and the reigning prince, who were running together. "Now I have got them!" Mukna must have thought. One more stride, and he would trample them to death!

But that instant the other bull elephant also ran close up to the two men—and hurled himself *between* Mukna and the two men.

Mukna's blow fell upon the bull elephant's side, and knocked him down. But Mukna tripped over him, and also fell. The two elephants rolled over and over upon the ground.

Meanwhile the young American and the reigning prince and all the other men, ran on to safety behind the trees.

When Mukna regained his feet, he realized that the men he had attempted to kill had escaped. And he also realized that now his punishment would be most terrible—first for the disobedience, then for the attempted murder. So in an instant he made up his mind to run away; he would escape to the jungle and become a wild elephant once more—even if he had to be a solitary wanderer in the jungle.

Sometimes in the wild West of America in [54] the past, men who had committed crimes would escape from the sheriff into the wilds and become outlaws. Mukna wanted to do just that. So he turned toward the trees on the side of the open space, to run away into the jungle.

But a most wonderful thing had happened. Without a word of command from anyone, all the other bull elephants had stepped to the gaps between the trees, each to the gap nearest him—as they would have done when they were wild elephants in a herd, to stop a criminal among them. And all of them were now facing Mukna.

Mukna turned to the right to find a way of escape to the jungle; but all the gaps on the right were guarded by bull elephants. Mukna turned to the left; but all the gaps on the left were guarded likewise. Mukna turned in all directions; but in all directions the gaps were guarded. He could not escape.

Then the elephant master recovered from his fright. He stepped out from behind the tree where he had hidden. For the first time he gave a command.

"March!" he cried to the elephants.

And the elephants marched toward Mukna. They came nearer and nearer, till they formed [55] a ring around Mukna near the middle of the open space. Mukna looked frantically this way and that way; but he saw a ring of elephants all round him, a dozen yards away; and the tusks of all were pointed toward him like a row of bayonets.

Then the elephant master and the royal party came and stood just outside the ring, at the back of the elephants.

The Trial of the Criminal Elephant—as in a Court of Law

There they held a trial, just as in a court of law. Mukna was accused of two crimes: first, disobedience; second, attempted murder. A man was appointed to defend him at the trial, just as in a court of law a criminal may have a lawyer to defend him.

The elephant master presided at the trial of Mukna. He was the judge.

When the trial began, Mukna's keeper first gave evidence; that is, he said that Mukna had disobeyed his order, not only once, but three times.

Then several other keepers came forward as witnesses, and gave evidence; that is, they said that they *saw* Mukna disobey the order. [56]

Then the man who was appointed to defend Mukna spoke for him; he was called the elephant counsel. The elephant counsel argued that Mukna must have been ill-treated to make him disobedient. So he questioned all the keepers. But all the keepers said that Mukna had not been ill-treated to make him disobedient.

"He may not have been ill-treated just that minute," the elephant counsel still argued. "But was he not ill-treated before? *An elephant has a long memory; he never forgets an injury, or an act of kindness.* An elephant has been known to remember both injury and kindness for more than twenty years. Then did not Mukna's keeper *ever* ill-treat him?"

But all the other men who were in charge of all the elephants gave evidence that Mukna's keeper had never ill-treated him; nor had anybody else ill-treated him—except that Mukna had been punished before for bad temper by being deprived of delicacies in his food. So Mukna had no true cause for disobeying the order that day.

Thus the charge of disobedience was proved against Mukna.

Then came the second crime of which Mukna was accused, namely, attempted murder. And [57] that was very quickly proved, as everybody there had just seen that crime.

So the elephant master, who was the judge, pronounced sentence of punishment on Mukna. Mukna was ordered to receive ten blows for the disobedience, and ten blows more for the attempted murder.

The Infliction of the Punishment

Now among the bull elephants forming the ring around Mukna was one who had huge tusks. So the elephant master ordered him to give Mukna the twenty blows. Of course the elephant could not *count* the number of blows he was to give. So the elephant master was to count for him, and tell him when to stop.

The elephant who had the huge tusks stepped into the ring, and tried to get behind Mukna, but Mukna turned around to prevent him from doing so. Then the elephant master ordered two other elephants to step into the ring. These two came and pointed their tusks at Mukna's ribs on each side. So Mukna could not turn. In defiance he held up his head, and curled his trunk tight before him.

"Hit me, if you like, but I won't give in!" he seemed to say. [58]

Five blows he took from the other elephant's tusks without flinching. But at the sixth blow he stumbled forward, and fell to the ground.

The elephant master stepped into the ring.

"Arise!" he commanded.

But Mukna would not rise.

Then the elephant master made a sign to the two bulls. They came to Mukna from each side, and prodded him in the ribs with their tusks. So Mukna was forced to stand up.

He steadied himself and received four more blows. Then at the next blow, which was the eleventh, he fell again.

"Arise!" the elephant master commanded.

Mukna again refused to arise. So the two bulls on the sides prodded him again, and forced him to arise.

This time Mukna stood only two more blows; then he fell again. The place where he was receiving the blows was now raw and bleeding. So the elephant master gave him a chance.

"Is it enough?" he asked.

But Mukna defiantly arose to his feet, without waiting to be prodded. And he defiantly held up his head and curled up his trunk.

"You may hit me as much as you like, but I won't give in!" he seemed to say. [59]

At the next blow, which was the fourteenth, Mukna again fell. He was getting weaker and weaker, and now he could not stand more than one blow at a time.

Seeing his weakness, the elephant master allowed him to lie there for five minutes.

Then he asked Mukna, "Is it *now* enough?"

Slowly, painfully, Mukna got up. He looked around with bleary, bloodshot eyes; he thought, "Can I not yet escape?"

But a row of tusks, like a row of bayonets, faced him on all sides.

Still he would not give in. With a fierce resolution he tried to curl up his trunk in defiance. He could not do so at once, but after an effort he succeeded.

"I won't give in, even if I die!" he seemed to say, though he was rocking unsteadily in growing weakness.

"Then we shall break your obstinate spirit!" the elephant master cried.

So Mukna received the next blow, which was the fifteenth. He fell. But after a while he rose again in defiance, and received the sixteenth blow. Then he fell in a heap. The side of his head hit the ground, and he rolled over. [60]

"Is it enough at last?" the elephant master asked. He waited.

Three times Mukna tried to raise his head in defiance, even as he lay on the ground; and three times he tried to curl up his trunk. His head went half-way up, and his trunk curled half-way. He lay on the ground just like that for a minute or two, his whole body quivering with pain and weakness.

Then perhaps the memory of all the kindnesses he had formerly received came back to his mind. Yes, an elephant never forgets an injury, but he never forgets a kindness either. Perhaps Mukna remembered at that moment all the petting he had received when he was a good elephant, all the sugar-canes and bananas and pancakes—and all the rewards for being gentle and docile and obedient. And now he realized that, instead of receiving these good things, he was receiving a most terrible punishment for being wicked, and for being *obstinate in wickedness*. How foolish he was!

He saw it all clearly in that moment, as he lay in shame and disgrace before all his comrades, all the other elephants. Then Mukna's head began to droop and droop; and his trunk began to unwind. The trunk hung loose and [61] limp before him; and his head sank lower and lower, till it lay humbly in the dust.

A low cry, almost like a moan, escaped his lips. It seemed to say, "I am sorry for being wicked and obstinate! I repent! Forgive me!"

Immediately the elephant master gave a sign. All the other elephants fell back. Their task was done. They returned to their usual work.

Then several of the keepers came with buckets of water, and bathed Mukna's wounds. Afterward they put on the wounds a poultice of herbs, to cure the wounds in due time.

So Mukna received only sixteen blows, instead of the twenty, because he repented of his crime.

"But if he had not repented?" you may ask.

Then he would have received the four remaining blows later on, when he was strong enough again to receive them. For the sentence of punishment must be carried out fully, like the sentence of a court of law, unless the criminal repents.

The Rogue Elephant

Among wild elephants in the jungle it sometimes happens that an elephant becomes so [62] wicked that he does not repent when he is being punished by the president of the herd. Then the president gives him as many blows as he can bear; that is, till he *cannot* rise from the ground. Then he is left there to recover by himself.

Sometimes such an elephant goes from bad to worse. For a few months his wounds may hurt him; and so he may be on his good behavior. But afterward, when the wounds have healed completely, he may commit a fresh crime. Then, of course, he is punished again. And now the place gets so sore and raw that it takes much longer to heal, and even then the place is full of scars.

If he should get unruly and commit a crime once more, would he be punished just the same? Yes, he would be. But I must tell you that a herd of elephants does not want a criminal among them. So after the third or fourth crime all the other elephants drive him out of the herd.

Then this very bad elephant meets a most awful fate. He becomes a solitary wanderer in the jungle. No other elephant will have anything to do with him. He is a *rogue elephant*.

"But could he not go to another part of the [63] jungle and join some other herd of elephants who don't *know* that he is a rogue?" you may ask.

He could. But those elephants would find out *at once* that he had been driven out of his own herd for being a rogue.

The Brand of the Rogue

How would they find that out at once? By seeing the scars of the wounds on the place where he had been repeatedly punished. Those scars are *the brand of the rogue elephant*.

So the new herd also would drive him out, for neither do they want a rogue among them.

Thus, no matter what herd the rogue elephant tried to join, he would be driven out.

Then he would be fated to roam the jungle by himself all his life — which is a most awful punishment. An outlaw among men has a similar fate, as he is shunned by all honest people.

A rogue elephant, being the *outlaw of the jungle*, does not live long. Just as an outlaw among men gets shot by the sheriff's men sooner or later, so also a rogue elephant gets shot by hunters. For, although the hunters must not shoot an ordinary wild elephant [64] that is a member of a herd, they may shoot at sight a rogue elephant that is roaming in solitude.

So, my dear children, remember that such a terrible fate comes to a rogue elephant who may have *begun* his downward path by just one act of disobedience or some other fault — and who obstinately persisted in his wickedness, and *would not repent*.

The Reward of Repentance

On the other hand, how much wiser it is to repent, even if one has been so foolish as to do wrong! Mukna committed the most terrible crime—he actually tried to kill people; and then he tried to run away into the jungle and perhaps become a rogue elephant. But afterward, when he was being punished, he repented of his crimes. So, what happened?

I shall tell you. Mukna was put on probation for a year; that is, the keepers watched him for a year to see if he would behave well. And for the whole year Mukna was on his best behavior; he was gentle and docile and obedient, and he did whatever he was ordered to do, even the hardest work. And he did that willingly, as if to prove that he had truly repented. [65]

Then those very princes whom he had tried to kill felt sure that Mukna had begun a new life, and would always be good in the future. So the princes took him back into favor.

And today Mukna wears a cloth-of-gold, with gold rings on his tusks, and he walks in a royal procession. Sometimes he carries grand people on his back, and sometimes children. And no elephant is more gentle and thoughtful with little children than he is. For he actually curls the end of his trunk near the ground for them to sit upon—and then he lifts them up to his back, three at a time!

[66]

CHAPTER V

Flesh-Eating Animals: the Felines, or the Cat Tribe

So far most of the animals I have described to you are vegetarians, that is, they eat vegetables of all kinds, for even leaves, herbs, and grass may be classed as vegetables. These animals are the elephant, the buffalo, the deer, the antelope, and others. The bear is the only animal I have so far described to you (in Book I) that eats both vegetables—that is, the roots of trees—and the flesh of other animals as well.

But now I shall describe to you quite a different class of animals, namely, animals that eat only meat. Among these animals the most important group is the Cat Tribe, or the *felines*, as they are sometimes called. They possess many of the qualities of the ordinary cat.

The principal felines are the tiger, the lion, the leopard, the puma, and the jaguar. All felines have a special kind of fangs, tongue, claws, and paws, which I shall now describe in detail. [67]

The Feline Has Strong Fangs

Besides the ordinary teeth, every feline has two pairs of strong fangs which look like big projecting teeth. One pair of fangs is placed on the upper jaw, pointing downward; they are wide apart. The other pair of fangs is placed in the lower jaw, pointing upward; they are not quite so far apart as the fangs of the upper jaw. Why? So that the animal can close its mouth comfortably without striking the lower fangs against the upper fangs.

These fangs are three to four inches long in a tiger or a lion; they are not quite so big in a leopard or other feline. The fangs of the tiger or the lion are so strong that he can hold down a heavy bullock by gripping it with his fangs. He can also drag the bullock along the ground by gripping it in that way, and can use the fangs to tear out a large piece of meat from the body of his prey.

When the tiger or the lion gets a piece of meat into his mouth, he uses the upper fangs to pierce the meat: that is, the meat lies on the ordinary teeth on the under jaw, and the two fangs of the upper jaw come down on the meat and cut it into two or three pieces. The tiger [68] or the lion could chew the meat a little more, with the help of his ordinary teeth, but he does not need to. Every animal of the Cat Tribe has a strong digestion; so the tiger or the lion merely cuts up the meat a few times with his fangs and then swallows it.

The Feline's Tongue is Rough

A feline's fangs, however, are too big to tear off *small* pieces of meat from a bone. So it uses its *tongue* to scrape off the small pieces of meat. That is the reason why a feline's tongue is very rough. So again you see, as I told you in Book I, that every animal has the gift it needs. If the feline did not have a rough tongue, it could not eat the small pieces of meat on a bone; and so a portion of its food would be wasted. No inhabitant of the jungle wastes food. It is only *we* who waste food.

The Feline's Claws are Retractile

The claws of every feline are *retractile*. That is, the claws can be *drawn in*, or sheathed, whenever the animal desires; also, the claws can be thrust out, whenever the animal desires to do that.

Why is it necessary for a feline to be able to [69] do both—to draw in its claws, and to thrust them out?

Because when the animal needs food, it must thrust out the claws to seize it. But in just running about in the jungle, it does not need to use its claws; so it draws them in. In fact, if it did not draw in its claws then, the claws would soon be worn out by rubbing against the ground. And even if the claws were growing all the time, they would be also wearing off all the time. So to keep the claws sharp for use only when the animal wants to seize something, it keeps the claws drawn in at other times.

Here I ought to tell you that a dog's claws are quite different from the claws of a feline, even from those of an ordinary cat. The cat's claws are of course retractile, as I have just described to you. But a dog's claws are *rigid*; that is, they are stiff and thrust out all the time. Why? Because the dog does not use its claws. It seizes its food with its mouth, not with its claws. It even defends itself with its mouth, that is, with its teeth.

But a feline uses its claws to seize its food, and even to defend itself. You may have noticed that even an ordinary cat defends itself with its claws. When a dog chases a [70] cat and corners it, the cat turns and defends itself with its claws.

Once upon a time, many, many hundred years ago, the dog did use its claws; they were then retractile. But the dog stopped using its claws; then they became rigid. The dog lost the power of drawing in its claws.

In our own bodies, if we do not use a particular gift for a long time, we lose the power of using that gift. When we are born, our left hand is just as good as our right hand. But because we do not use the left hand much in doing things, we lose the power of using it quite as well as we use the right hand. Little boys and girls should practice using the left hand. Then if by some accident the right hand is lost, they would not be quite helpless.

As for the felines, they retain the full power of their claws by constant use. So, because the claws are very useful, every feline takes care of its claws,—especially the tiger. Why, *the tiger cleans his claws every day*! In the jungle there are many trees that have a soft bark. So the tiger goes to one of these trees every day, and digs his claws into the bark. Then he draws his claws sideways along the bark, and that cleans out the claws. The [71] tigress also cleans her claws every day in the same manner.

Some little boys and girls do not clean their nails every day. Then sometimes a piece of dirt gets in under a nail and causes a sore. But the tiger and tigress are wiser. If part of a piece of meat that they have torn up were to remain under a claw, it would fester and cause a sore. So the tiger and tigress clean their claws every day.

The Feline Has Padded Paws

The paws of every feline have also a special quality. The under part of each paw is thickly padded with powerful muscles. That gives the feline three advantages.

First advantage: it enables the feline to *stalk* its prey. That is, the feline can creep up to its prey quite silently. As its paws are padded, they make no sound on the ground—just as your footfall makes no sound when you wear rubbers over your shoes.

Second advantage: the padded paw enables the feline to strike down its prey with a severe blow. When it wants to strike down its prey, the feline hardens the muscles under its paw; the blow of its paw is then something like that [72] of a hammer. A tiger has often been known to smash the skull of a buffalo with a single blow of its paw.

Third advantage: the padded paws enable a feline to leap farther. After a feline has crept up as near to its prey as it can, it has still to leap upon its prey to seize it. Then the muscles under the paws act like springs, and enable the feline to give a big leap. Even in running, the muscles act somewhat like springs. You must have noticed that, in running, a dog *gallops*, but a cat *bounds*. That is, the dog moves its legs very quickly, but each space of ground it covers is not very long. A cat moves its legs not quite so quickly, but the space of ground it covers at each bound is much longer. The cat and all felines can give a bigger bound because of the muscles under their paws.

Having told you all the qualities common to animals of the Cat Tribe, I shall now describe some of these animals in detail.

[73]

CHAPTER VI

The Tiger

The tiger lives in most of the countries along the south coast of Asia, that is, all the way from Persia to China. Some tigers are also found in the northern countries of Asia, such as Siberia; but there are very few of them there. And, of course, these few tigers in the cold northern countries of Asia are a little different from those in the hot southern countries. For the tigers in the cold countries have thick fur on their skin, and a layer of fat under their skin—just to keep them warm. So they are too fat to be as muscular and active as the slim and lithe tigers that live in the hot countries in the south of Asia.

Now please remember one thing more about the dwelling place of the tiger: *there is no tiger in Africa*. Even clever people do not always know that. When ex-President Roosevelt went on a hunting trip to Africa a few years ago, he shot many wild and ferocious animals there, and some newspapers said that he had shot several tigers. [74]

That was a mistake. The animals that he shot were leopards, not tigers. You can at once tell the difference between a leopard and a tiger: a leopard is *spotted*, but a tiger is *striped*. I shall tell you all about that presently.

Even as regards the habits and character of the tiger, people often make mistakes. There is no animal that has been so much abused as the tiger. Most people call the tiger a "cruel" and "bloodthirsty" animal.

But that is not true. By "bloodthirsty" people usually mean that the tiger kills his prey for the mere sake of killing, and that he kills more animals than he can eat, just for the mere fun of killing.

That is not true. A tiger is not really "bloodthirsty" in that way, as I shall explain to you presently. A tiger never kills for the mere fun of killing. Some men and some naughty boys do that! They think it great sport to kill harmless wild animals, which they cannot possibly eat or use in any way; and some naughty boys kill frogs and lizards and other small animals, just for the mere "fun" of killing, as they call it.

[75]
[76]

Tiger

A tiger never does that—and he is supposed [77] to be the worst animal of all! For one thing, a tiger is not such a fool as to kill his prey for the mere sake of killing. Men formerly ate the flesh of the American bison, or buffalo, as it was generally called. But then they killed off whole herds of these buffaloes. So now there are no more buffaloes left for food in those places. A tiger is wiser. He does not destroy his own food supply needlessly.

People are also wrong when they say that a tiger is "cruel," and that he tortures his prey before killing it outright. That is not true of the tiger. In fact, hardly any animal is *needlessly* cruel, as some men and naughty boys are—for instance, naughty boys who torture frogs and lizards and then kill them.

It is true that a *tigress* does worry her prey before killing it. But why does she do so? Simply to teach her cubs how to catch and kill the prey, so as to provide food for themselves when they grow up. I shall explain that fully presently. So please remember this once for all: hardly any animal is *needlessly* cruel or bloodthirsty.

"But a cat does worry a mouse, before killing it," you may object. "Is not that needless cruelty?" [78]

That seems quite true. But there is a reason for it: the cat first began to do that to teach her kittens how to catch mice, when she was a wild animal in the fields. Once upon a time the cat was a wild animal, but now people have tamed it into a domestic animal. So the cat still retains some of its wild habits.

But you will understand all that when I tell you more fully about the tiger, which is the largest and strongest animal of the Cat Tribe.

The Life History of the Tiger Family

I shall describe to you the actual life of a tiger family in the jungle. A tiger family consists of the father, the mother, and from two to four cubs. Three is the usual number of children that a tiger and tigress have.

When the cubs are only a few days old, they are quite helpless. So the mother stays with them in the den, while the father goes in search of food. The den is usually a hollow under a large tree.

If the father tiger catches a prey which he can carry, such as a deer, he brings it home with him. Then he and the tigress eat it together.

But if the prey is too large to carry, such as a bullock or a buffalo? Then the tiger first eats [79] a good portion right after catching it. Then he comes home to the den and sends out the tigress to eat her share, while he stays home in the den and takes care of the cubs.

But here is something for you to think of. In sending the tigress out to eat her share of the prey, the tiger must *tell* her where the prey is lying; otherwise she might go the wrong way. Why? Because the prey might be lying a mile or more from the den, so that she could not possibly trace it merely by its *scent*. And the prey might have been caught in any direction, especially if the tiger had to chase it or stalk it for a long distance. So nobody could tell beforehand in what direction a tiger might catch its prey.

The tigress could not merely follow the tiger's *paw marks* to get to the prey, as the tiger may have gone out several times that day or

55

the day before; and so there would be several lines of paw marks, and she would have to search very long by following all the paw marks in turn. Yet she always takes the right direction, and gets to the prey quickly. Hunters in the jungle have found that out. How does she do it?

The only way to explain it is this—the tiger [80] *tells* her where the prey has been caught and is now lying. That is what hunters believe from the actual facts they have observed. Then that shows that animals have a method of communicating with one another. Of course they do not use *our* words. They must have words or sounds, or even signs, of their own.

Now I shall go on with the tiger family. The cubs, of course, drink their mother's milk. They do that till they are three months old.

But meanwhile, when they are six weeks old, they can walk and trot. They are then very playful, and they leap and gambol and tumble over one another.

They are then able to go about with their father and mother for a short distance. So if food gets scarce for the tiger and tigress, they leave their old den altogether, and go to live elsewhere in the jungle where food may be more plentiful.

In this house-moving the cubs can trot behind their father and mother for a mile or two. Then, for fear of tiring the cubs, the tiger and tigress scoop a hollow under a tree, and place them there. The tiger and tigress go on ahead till they find the new home. Then they come back to fetch the cubs. [81]

If the cubs are now two months old, the father and mother need have no fear in leaving them for a few hours. So in their new home the tigress may go hunting with the tiger every day.

If food gets very scarce, the tiger goes out alone for a long distance for two or three days at a time. In his absence, the tigress makes a short trip from time to time in another direction, in case any other kind of food may by chance be found there.

Tigers prefer to eat deer or antelope, just as you may prefer to eat roast turkey. But if tigers cannot get deer or antelope, they have to catch a bullock or a buffalo—which is just plain beef! As even that may be scarce, tigers have to be satisfied with the wild pigs, which are plentiful in the jungle,—that is, just pork! As a change now and again, they may have mutton, because there are also wild sheep and wild goats in the jungle.

But when the tiger and tigress are both away from the den in search of food, are the cubs quite safe in the den?

They usually are, after they are two months old, when they are as big as house dogs; and, until then, either the tiger or the tigress stays [82] with them all the time. When the cubs are two months old, they may stay by themselves in the den; then a wolf or a hyena may perhaps come to the den, and try to kill one of the cubs; but all the cubs would stand together, facing the enemy, and would defend themselves.

They would change at once from being playful like kittens; they would become little tigers in their nature. And woe to the wolf or hyena when the mother returns! She would know at once by the cubs' actions that they had been annoyed. Then the tigress would track down the intruder and kill it.

At the age of three months the cubs can eat meat, but they cannot chew it as yet, as their teeth are only beginning to grow. So the mother chews the meat for them. If she or the tiger has caught a deer, she chooses the tenderest part of the meat, and chews it into mincemeat. Then she puts a little of it into the mouth of each cub. She does that several times, till the cubs have had enough to eat.

When the cubs are four months old, their teeth have grown enough to enable them to feed entirely by themselves—but only on very tender meat. [83]

The Tiger's Family Dinner

It is very interesting to watch a tiger family having their dinner. I may remind you again that some hunters who go into the jungle sometimes hide in trees and watch the family life of different animals. So this is what they have observed at the tiger's family dinner.

Suppose that the tiger has brought home a blue deer, which is a great delicacy among tigers. He drops the blue deer in front of the den. He and the tigress lie down and watch the cubs, who eat first.

The tiger or the tigress will not tell the cubs which portion of the deer is the tenderest; they must find that out by themselves. That will be their *first lesson* in life.

So the tiger and tigress keep aside, and see what their children do. One of the cubs makes a sudden grab at a leg of the deer, and tries to tear out a mouthful; but to its disgust the cub finds that it cannot bite the leg of the deer at all. I suppose then the father tiger gives a sort of wink at the mother tigress; at any rate, the tiger and tigress just look on, and say nothing.

Then another cub has a bite; perhaps it [84] tries the back of the deer's neck. But this cub also finds to its disgust that its teeth will not go through the meat there.

In this way the cubs jump about the deer, and try to bite it in different parts. They get more and more disgusted; but still the father tiger and mother tigress say nothing.

Then at last one of the cubs dives in, and makes a grab at the *throat* of the deer—and to its delight it finds that the meat there is quite tender, and that it can tear out a piece very easily. Of course that cub eats it quite greedily, and then has several more mouthfuls. But then—

"You have had enough!" says its father. "Give Brother and Sister a chance!" Of course the father tiger does not say that in *our* words; and he need not say it in any kind of words. He just comes to that cub and tumbles it over with a gentle pat.

Then the other cubs come to the throat of the deer, and have their dinner also. As there is not sufficient meat there to satisfy all of them, they soon find that the under part of the deer is also tender enough for them to eat.

The father tiger and mother tigress eat last, [85] when all the children are satisfied. The tiger and tigress of course can eat any kind of meat, so they eat the legs of the deer. And if it is a deer of ordinary size, the tiger family finishes it altogether at one meal!

So you see how kind the tiger and tigress are to their children. Suppose that among us there was a family of five people, father and mother, and three children; and suppose they were having a turkey dinner. Then if the father and mother were as kind to their children as the tiger and tigress are, they would give to their children the breast and all the nicest titbits of the turkey—and after that the father and mother would eat what remained of the turkey.

That shows that a tiger is an affectionate father, whatever faults he may have. Among animals, the mothers, of course, are nearly always affectionate to their children; but very often the fathers are not. In fact, among some kinds of animals in the jungle, the fathers do not care much for their children; they desert them.

But the tiger is different; he is usually a good father. That is an important thing to remember. It shows that even if an animal is [86] supposed to be very bad generally, it may yet have some special virtues of its own. That is a lesson for us. We may know people who are supposed to be bad; but even then we should try to find out if they have some good quality.

[87]

CHAPTER VII

The Tiger Cubs' Lessons

Do tiger children have lessons? Of course they have! Almost all animal children have. You will remember the lessons in Book I which the elephant child had to learn. In the same manner other animal children must learn how to make a living in the jungle, and also how to avoid dangers.

Among tiger children, their lessons begin even when the father and mother are providing them with the food; for, as I have just told you, the children must learn at least which part of the meat to eat, and which not to eat.

But the most important thing they have to learn is how to catch the prey, and how to kill it for themselves — that is, how to provide their own food. Their parents teach them to do that gradually from time to time, in many lessons. [88]

Tiger Cubs Learn to Kill Prey, After their Parents have Caught It

As they are not yet big enough to *catch* the prey, they are first taught how to *kill* the prey, after their father or mother has caught it alive for them. And that is another wonder of the jungle, and another good quality of the tiger. If the tiger catches a deer, even the largest kind of deer, he could kill it at one blow, so as to eat it at once. But if the tiger is the father of a young family, he thinks of his family all the time; he remembers that he must not only provide his young children with food, but he must also *teach them their lessons*.

So when he finds a big red stag, he jumps upon it, but he does not kill it outright. Instead, he merely breaks its hind legs, so that the stag cannot run away. Then he calls the cubs and the mother tigress. The tiger and tigress stand aside, and tell their children to kill the stag. They will not at first show the children how to do it. The children must try first to find that out for themselves.

So the cubs first prowl around the stag, and try to seize it anywhere. But the cubs cannot get their teeth deep enough into the stag's [89] body; and as the stag is still alive, it shakes them off. The cubs try to seize the stag at other parts of its body, but each time they fail to hold on; instead, the stag shakes them off. And if the cubs dare to come in front of the stag, the stag can still use its antlers to drive them off.

Then how can the tiger cubs manage to seize the prey at all with their teeth? Well, one of the cubs may remember the very first lesson it had several weeks before: that was to eat the *throat* of the prey, because it was the softest part—as I have already described to you. So it remembers that the throat is the softest part.

Then that cub comes to the side of the stag, makes a sudden plunge downward, and seizes it by the throat. Even then the stag tries to shake off the cub—but the other two cubs then come to their brother's help; they also seize the stag by the throat, one from each side.

Thus the three cubs begin to *worry* the prey, that is, they shake it, and pull it, while their father and mother watch them. The prey holds up its head and struggles, but gets more and more exhausted with the weight of the three cubs. At last the prey is unable to hold [90] up its head any more. Its head sinks to the ground. Then the three cubs kill it easily.

Tiger Cubs Take Part in Hunt to Catch Prey

When the cubs are six months old, they can take part in the actual hunt for the prey. So they go into the jungle with their father and mother. When they sight the prey, the cubs stay a little behind, while the father and mother stalk the prey.

Suppose the prey is an antelope. You will remember what I told you in Book I, that an antelope looks like a deer; but it is a little different from a deer, because an antelope has horns, and a deer has antlers. Well, the tiger creeps around to the side, then more and more around, till he gets behind the antelope.

Meanwhile the tigress creeps around the opposite way. So when the tiger makes a sudden jump at the antelope, and the antelope tries to run away in either direction, the tiger or the tigress is there to catch it. And meanwhile the cubs also have crept nearer and nearer, hiding behind shrubs and bushes. They can take part in catching the prey by preventing it from escaping in their direction. [91]

Tiger Cubs Learn to Catch Prey by Themselves

"But when do the tiger cubs actually learn to *catch* the prey?" you may ask.

Well, that takes a little longer to learn. For when the cubs have learned to catch different kinds of prey—wild pigs, wild sheep, wild goats, deer, antelope, cattle—their education is almost finished, just as in the case of a boy who has learned to earn his living in several different ways. So it takes the tiger cubs at least the next four months, from the age of six months to ten months, to learn to catch different kinds of prey, as I shall now describe to you.

In the beginning the cubs learn by example; that is, they watch and see how their father or mother catches the prey. Some kinds of prey are very easy to catch, such as wild pigs or wild sheep, as they cannot run fast, and are also very stupid. A tiger can just rush at a wild pig or a wild sheep, and catch it. So the cubs soon learn to do the same. And as I have already told you that wild pigs and wild sheep are the usual food of tigers, the cubs soon learn to earn their *ordinary* living. [92]

But then they have to learn a little more difficult lesson—to catch animals which are not so easily caught; and these animals supply them with a more tasty kind of food than just pork or mutton. These animals may be divided into two classes.

First, the prey may be weak, but it can run fast—even faster than the tiger. The deer and the antelope belong to this class.

The second class of prey is just the opposite; it is strong, but it cannot run fast—at least, not as fast as the tiger. Buffaloes, bullocks, and all kinds of cattle belong to this class.

In catching these two different kinds of prey, the tiger or the tigress uses different methods. First I shall describe to you how a tiger catches an animal of the first kind, that is, an animal that is weak, but which can run faster than the tiger, such as a deer.

Can you think how the tiger does that? He cannot *chase* the deer and run it down in the open country, because the deer can run faster than the tiger.

"The tiger can hide in the tall grass near a river, and wait for a deer to come to drink," you may say. "Then the tiger can jump on it." [93]

That is quite true. And the black stripes on the tiger's yellow body make him appear very much like the tall grass where he is hiding. So the deer does not notice the tiger, and it often comes quite close to the tiger to drink—and then the tiger jumps on it and catches it.

But a tiger may also catch a deer by *stalking* it. If he sees a deer browsing at a distance, he tries to creep quietly toward the deer. He hides behind bushes and thickets every few minutes, then he creeps on again toward the deer. He does that very cleverly. If the deer is bent on feeding, the tiger creeps on for a few yards. But if for a moment the deer stops feeding, the tiger hides at once.

In this manner the tiger sometimes creeps to within a few yards of the deer. Then he gives a sudden spring and falls on the deer. If he cannot approach the deer near enough to fall on it with just a spring, he first makes a swift rush and *then* he gives the spring.

When a tiger or a tigress is teaching the cubs to stalk a prey in that manner, the cubs of course stay in the rear and hide behind a bush, and from there they watch. So they see how their father or mother stalks the prey—as I [94] have just described to you. Of course, they have to watch their father or mother several times before they learn that lesson fully.

Now I shall tell you how a tiger catches prey of the other kind—that is, an animal that is strong, but which cannot run fast, such as a bullock. The tiger comes toward the prey from the side or from the back, but never from the front. Why? Because the prey has horns, and if the tiger tried to attack it from the front, the prey would gore the tiger with its horns and perhaps kill the tiger.

So the tiger creeps toward the prey from the side or the back. As the prey cannot run very fast, the tiger does not trouble to stalk it all the way. Instead, the tiger creeps up to within a hundred yards of the prey; then he gives a number of quick rushes, till he reaches the prey. And he is always careful to reach the prey from the side or the back.

"But if the prey turns in time and faces the tiger with its horns?" you may ask.

Then the tiger turns also. He dodges from side to side. A tiger can always turn faster than any horned cattle. A tiger may even come to within a few yards of the prey, and jump clear over it! Then on landing on the [95] ground, the tiger can turn at once and reach the prey from the side. Then he gives a quick blow with his paw on the neck of the prey. One blow is usually enough to stun the prey and knock it down.

Sometimes the prey is so frightened when it first sees the tiger, that it does not try to face the tiger with its horns at all. Instead, the prey stands trembling with terror, and lets the tiger come right up to it from the side. Then the tiger gets up on his hind legs, places one paw on the prey's shoulder, and with the other paw he gives a terrific blow on its neck.

But if the prey is not too frightened, and it struggles when the tiger is trying to strike it, then the tiger uses a different method. He plunges downward and seizes the prey from underneath by the *throat*. He plants his hind legs firmly on the ground, a little bit away from the side of the prey. In that way he gets a little more "leverage," as it is called.

You have seen a man tilt a heavy box over on its side by placing a crowbar under it, then lifting up the crowbar. Well, the tiger acts somewhat like that. While still holding the prey by its throat in his jaws, he gives a sudden jerk upward with his head. In that [96] way the prey loses its balance and topples over on its side, just like the box.

When the tiger or the tigress is teaching the cubs to catch horned cattle in these different ways, the cubs of course stay a little behind and watch how their father or mother does it.

So in every case, as you will understand, the tiger cubs have to learn from their parents how to get their living in the jungle.

[97]

CHAPTER VIII

The Tigress Mother's Special Duties

So far I have described to you how the tiger cubs learn the lessons of the jungle from their father and mother.

But sometimes they have to learn some of their lessons from their mother alone. Food may be scarce in that part of the jungle. A tiger family eats so much that even if they catch a large wild pig or a deer every day, it will hardly provide more than a single meal for a tiger, a tigress, and two or three growing cubs.

And as they do not usually catch prey every day, the family eats only about two or three times a week. When the cubs are from six to ten months old and need more and more food, one prey at a time is not enough to provide for the whole family — if they all live together. So it is better for the family that the father should go away and catch his own food, while the mother catches food for herself and the cubs.

But before going to earn his living elsewhere, the tiger takes his family to the *easiest* hunting [98] ground there is near their jungle, where there is at least some kind of prey to catch. Then the tiger himself goes to a more difficult hunting ground. So even in that a tiger is kind to his family, and he does the best he can for them.

At first he returns to the family every few days; I suppose he does that to see how they are getting along in his absence. By that time the cubs have learned most of their lessons, and the mother tigress continues the lessons during the tiger's absence.

But after the cubs are ten months old, they have learned all their lessons; they only need to *practice* what they have learned. As they can do that with their mother, they do not need their father any more. So the tiger then goes on his travels to distant parts.

As the cubs practice with their mother the different ways of catching and killing the prey, she must provide them with many chances of doing so. It is then that she helps the cubs to kill more animals than they can eat. That is why people give the tiger a bad name and call him a "bloodthirsty" animal. It is not he at all, but the tigress mother. And she helps to kill a large number of animals only at [99] this time—when she must provide her cubs with the chance of practicing their lessons.

The tiger cubs do not need even their mother when they are two years old. By that time they are quite able to get their own living by catching every kind of prey. But still they usually stay on with their mother for about six months more. Then they leave their mother, and roam the jungle alone, each cub separately.

But each cub still continues to grow in *size* till the age of four years. A male tiger may even grow in *strength* till he is six years old.

But you may want to know if a tiger family ever meet again after they have all separated. That may sometimes happen. It may be in the dry season, when nearly all the water in the jungle is dried up. Then by some wonderful instinct *all* the animals in the different parts of that dry region know that there may be one place where there is water. So a general migration begins toward that place; that is, *all* the animals begin to travel to that place with their families.

These animals may start from different places a hundred miles apart, and yet after a few days they will get to that same Water [100] Hole. Of course they do not all reach it on the same day; but many of the animals stay near there for a few days, till the rain comes and there is water in other places. So it does happen that a tiger family may meet again at the Water Hole, and then there is a happy reunion among them.

The Truce of the Water Hole

But the tiger family must not kill a prey at the Water Hole. And all other flesh-eating animals—lions and leopards, and wolves and hyenas—must also abstain from killing prey there. Hundreds of pigs and sheep and deer may have come to drink at the Water Hole—- and every flesh-eating animal must abstain from killing any one of the pigs or sheep or deer.

This "Truce of the Water Hole" is one of the greatest wonders of the jungle. It means that in other parts of the jungle there may be a kind of war, because flesh-eating animals may kill and eat their prey, but when all the different animals meet to quench their thirst at the Water Hole, there must be no war—no killing, no fighting. There must be peace at that place while the different animals are there. [101]

At the Water Hole the tiger and the lamb may drink together in peace; and hungry as the tiger may be, he must not hurt the lamb. And the wonder of it is that the tiger knows that law, and always keeps it. Likewise all other flesh-eating animals always keep that law; they never hurt even the weakest and most timid animal at the Water Hole.

They all feel that they have come there for a greater need than *hunger*—they have come there to quench their *thirst*; and the pain of thirst is greater than the pain of hunger.

They feel that the pain of thirst is common to them all; that is, they all suffer from that pain. Different animals *eat* different things; but they must all *drink water*. And in that fellow feeling there is peace among them all.

My dear children, let me impress this upon your minds, while you are still young. When you grow up, you may sometimes be tempted to doubt that an all-merciful Providence watches over us. Then remember these wonders of the jungle that I have described to you. And remember especially the Water Hole, where all animals are like brothers, where even the tiger and the lamb drink and lie down together in peace.

[102]

CHAPTER IX

The Special Qualities of Tiger and Tigress

Now I am going to tell you a few more things about the tiger, from which you will realize what a wonderful animal he is.

First, the tiger's *size*. The finest specimen of the tiger is the Royal Bengal tiger. Such a tiger, when full grown, is sometimes seven feet long, without including the tail; the tail is usually half as long as the body. The tigress is slightly smaller.

In height a Bengal tiger often measures three and a half feet from the shoulder to the ground; so his head would be more than four feet from the ground. Hence, if you take his length into account, you will understand that the tiger is really the largest feline or animal of the Cat Tribe.

I do not think that you have often seen a really large tiger in the zoo. Most of the tigers in a zoo were caught as cubs; that is, the mother or the father was shot by hunters, and the cubs were captured alive.

Now, just think. If a human child were locked up in a room all his life, without any [103] exercise, then he would be very stunted and small, even when he had reached the age of a man. So a tiger cub, brought up in a cage all its life, never grows to its proper size. For this reason most of the tigers in a zoo are much smaller than those tigers that grow up in the jungle.

The most wonderful thing about the tiger is his strength; he is the strongest animal of the Cat Tribe. That is proved by the way in which he carries his prey. If the prey be a deer or a man, he seizes the prey in his jaws by the middle of the body—just as a cat seizes a mouse! And the tiger carries such a prey in that manner to his den, which may be more than a mile away.

But a heavy animal, such as a cow, he carries in a different manner. Yes, a tiger *carries* away a cow; he does not merely *drag* it along the ground, as a lion does. This is the way the tiger carries a cow, after killing it:

He first seizes the cow in his jaws by the back of its neck. Then he rears up on his hind legs and swings the cow over his shoulder on to his back—just as a man swings a loaded sack on to his back. Then the tiger stands on all four legs again, and trots along with his burden. [104] Of course, he still holds the neck of the cow in his jaws, just as the man carrying the sack holds the upper end of the sack in his hand.

I shall now finish with tigers by telling you three stories,—true stories, of course. From these stories you will understand that tigers and tigresses sometimes have the same kind of feelings that *we* have.

Both Tiger and Tigress Defend their Cubs

I have told you that in a tiger family, when the cubs are very young, they must be guarded all the time by either their father or their mother. One day it happened that a tiger had killed a bullock. As he could not carry it to his den, he first ate enough of the bullock to satisfy his hunger. Then he came home to his den, and sent the tigress out to eat her share, while he guarded their two cubs in the den.

But three English officers had gone hunting in the jungle, each of them on an elephant; and it so happened that they came toward the tiger's den.

The three hunters saw the tiger and the two cubs he was guarding. The hunters knew that if they killed the tiger they could catch the two cubs alive. So they fired their guns at [105] once at the tiger; and as they were then only about a hundred yards away, they all hit the tiger.

Now, if the tiger had not had the cubs to defend, there would not have been much of a fight. Why? Because, as I shall tell you later, it usually takes much more than three hunters on three elephants to hunt one tiger. Each of the three wounds the tiger got might have killed or disabled any other wild animal; but instead, the three wounds together only made the tiger furious.

If he had been alone, he would have come like a flash of lightning at the nearest elephant, leaped upon its back, and killed the hunter on it—before the hunter could shoot again. Of course, the other two hunters could then kill the tiger; but the tiger would at least have killed *one* of the hunters.

That is exactly what the tiger would have done, if he had been alone. But the tiger had his children to defend. He must try to guard them as well as he could. So he just took one of the cubs in his mouth—as you have seen a cat take up her kitten—and leaped with the cub over a thicket and hid the cub there.

Then he leaped back to the den to take [106] away the second cub. That gave the three hunters enough time to load and take aim again. So all three of the hunters fired at the tiger again, just as he was lifting up the second cub; and the bullets went through his heart. If he had been any other animal, he would have dropped dead right there. But a tiger lives about three seconds after he *ought* to be dead; and in those three seconds he can give just one leap and kill anything.

But the hunters were beyond his reach. So he gave that one leap toward them, and tore up the ground instead, as he could not tear up the men; then he agreed to lie down and be truly dead.

The three hunters got down from their elephants and came to the den. They found that one of the last bullets had passed right through the tiger's body, and had killed the cub he was trying to carry to safety. The hunters were sorry that the cub had been killed. So they searched for the first cub, which the tiger had hidden behind the thicket. They found the cub and took it with them.

[107]
[108]

Tiger Protecting his Cub

The hunters mounted their elephants and came back to their tent, where they had been staying. They put a dog's steel collar around [109] the neck of the cub, and tied him up to the tent post by a chain. The cub was so frightened and helpless that it lay down on the ground and was very quiet. The three men sat down in the tent and chatted for a while.

Suddenly they heard a terrible roar outside. They snatched up their guns, but they could not tell from which side the roar came — just as when you hear a terrible clap of thunder close by, you cannot tell from which side the thunder comes. And hearing this roar, the cub jumped up and yelped in answer; and he tugged at his chain furiously. He had become a little tiger in his nature.

Suddenly a huge yellow form shot into the tent. It was a tigress. She seized the cub's collar in her mouth, and snapped the chain with a tug, like a piece of thread. The next second she leaped out of the tent with the cub, and vanished. And the three men had not had time to aim a gun. None of them really wished to.

Yes, she was the mother of the cub. When she had returned home from dinner, she had found her home broken up — her husband killed, one of her children killed, and the other child stolen. So, all that she could do was [110] to regain her lost child by tracing it by its scent.

This she did. She regained her cub even by facing the same guns that had killed her husband. For a tigress mother, like any other mother in the jungle, will face death to save her child.

The Tiger Family's Lost Dinner

Now I shall tell you another true story. It will show you what sort of a husband and father in everyday life a tiger is.

Near a jungle there was a river. At a special place in the river there was a bend. It was a good place for fishing, as the water there had plenty of fish.

One afternoon two men went to fish there with fishing rods. As there was a jungle about a mile from the place, the men took their guns with them, in case any wild animals came from the jungle to attack them.

After a time one of the men hooked a fish. It must have been a big fish, as it tugged at the line furiously. The man who had hooked the fish had to run along the bank of the river to *play* the fish, while his friend kept shouting to him to advise him what to do. In this way [111] both the men were busy, and forgot to think of anything else.

Suddenly they saw a flash of yellow. It came straight from a bush toward the man who had hooked the fish. It was a tiger!

The tiger must have stalked the two men silently from the jungle; and in that way he must have crept up to the bush, while the two men were busy trying to land the fish.

The tiger gave a rush and a leap, and fell upon the man who had hooked the fish. He grabbed the man and leaped back with him into the bush, before the other man could snatch up his gun and take aim to save his friend.

Now you will remember what I told you: that a tiger carries a man in his jaws just as a cat carries a mouse; that is, the tiger holds the man by the middle of his body, about the waist.

Luckily the man was wearing a waistband of thick cloth; so the tiger's fangs did not hurt the man very severely, as the fangs happened to bite the thick waistband. But still the man had been stunned by the shock when the tiger had leaped upon him. And the tiger thought that he had killed the man [112] outright. That was very lucky for the man—as you will understand presently.

The man regained his senses while the tiger was still carrying him. He knew at once that he was in the jaws of a tiger. That is perhaps the most terrible danger for a man to be in. Few men have ever been in the very jaws of a tiger in the heart of the jungle—and yet have escaped.

The man knew that, and so he was terribly frightened. But life is so precious that one must never despair of saving his life. If you are in the most terrible danger, *you must never give up hope*. You must try to find some way of escape.

So the man began thinking, even while the tiger was carrying him. He made up his mind at once. He must pretend to be dead. So he did not move or make the least bit of sound. Even then he did not see how he could escape, as the tiger would soon start *eating* him! But still he would not despair.

The tiger carried the man to his den in the jungle. The den was just a hollow in the ground under a large tree. The tiger dumped the man into the hollow. The man thought his end had now come. He could not escape [113] from right in front of the tiger's eyes. And he thought that the tiger would start eating him at once. Even though he was really alive, the tiger would eat him just the same.

But, to his surprise, the tiger did not start eating him at once. Instead, the tiger looked around, and gave a purr, and then a growl. What did that mean? The man could not tell.

Then the tiger just flung upon the man some of the sand from the side of the hollow. The man understood *that*: the tiger was trying to hide or *cache* his food—as some wild animals do.

But luckily the tiger only flung the sand loosely over the man, just enough to cover him; he did not quite bury the man; or else the man might have been smothered. Then the tiger ran off into the jungle.

The man was puzzled to know what the tiger meant by that. But you may be sure the man did not wait to work out the puzzle in his mind. Instead, he jumped up from the hollow. Here was his chance to escape!

But he was afraid to run far; for the tiger might return at any moment and catch him again. So the man just climbed up the tree [114] under which the den was. And he went up the tree as high as he could, and hid himself among the leaves.

After a while he heard a sound below, at a little distance. He looked down and saw the tiger returning. But now there was a tigress with him, and two cubs.

Then the man understood the puzzle. When the tiger had brought home the dinner, he had found that his wife and children were out. So he waited a while; and as they still did not come home, he first looked around for them, and then he gave a loud *call* to his family to come to dinner. That was the purr and growl he gave.

As they still did not come home, the tiger just hid the dinner to keep it safe, and then he went out to *fetch* his family home to dinner.

But when he did fetch them, the dinner had run away! Then the tiger family set up such a wail and lament over the lost dinner!

"I felt quite sorry for them," said the man up in the tree, afterward. "They kept up the wailing and growling and lamenting for a long time. Only, as it was *I* who was to have been the tigers' dinner, I wasn't so very sorry that the dinner had escaped!" [115]

Meanwhile, the other man who had been fishing with him had run to the nearest village. The villagers got together a herd of bull buffaloes, and started tracking the tiger by the paw marks he had made on the ground. In this way the villagers brought the bull buffaloes to the tiger's den.

The bull buffaloes soon drove away the tiger family. The villagers expected to see only the man's bones or half-eaten body. But still they had come to make quite sure of the man's fate.

What was their delight, then, to hear a shout, as soon as the tiger family had been driven away! The shout came from the tree. It was from the man who had been carried away by the tiger. You may be quite sure that he was very glad to climb down and go home with the villagers.

Now, my dear children, I have told you this story—and it is a true story—for two reasons. First, it shows you that you must never give up hope, even in the worst danger. If a man can escape from the very jaws of a tiger in the heart of the jungle, he may be able to escape from other dangers.

The second thing I want you to learn is that, [116] bad as he is supposed to be, a tiger is really a good husband and a good father, even in ordinary everyday life. When he had earned the dinner, and had brought it home, he found that his family was out. He might have started eating the dinner himself. Instead, he waited for his family to return, then he called out to them, and then he went to fetch them—without eating a bite himself. How many *men* would do that?

The Tiger as a Heroic Husband

Now I shall tell you another true story, which will show you in a different manner what a wonderful animal the tiger is. It is the story of a great tiger hunt.

A few years ago Prince Henry of Orleans was one of the greatest hunters in the world. He had hunted lions and wild elephants in Africa, and also other big wild animals. Then he went to India, hoping to hunt tigers.

There he was the guest of a rajah, that is, a sort of king. So the rajah arranged a tiger hunt for Prince Henry. In a jungle near by there were many wild animals. On the north side of the jungle there was a shallow ravine, only about ten feet deep, and as wide as a street. [117] The ravine started from the jungle and went northward. Beyond the jungle the ravine ran for only about a hundred yards; beyond that the ground was level again.

It was right there on the level ground, in front of the ravine, that the rajah placed the hunters. The hunters were mounted on thirty elephants, two hunters on each elephant; so there were sixty hunters altogether. The two hunters on each elephant sat in a kind of big box, called a *howdah*. The box was tied fast on the elephant's back with strong ropes passed all round the elephant.

Meanwhile about a thousand men started toward the jungle from the fields on the south side of the jungle. As they came near the jungle, the men made a loud noise with drums. So all the timid animals in the jungle took fright and began to run away. These timid animals were the deer, the antelope, the wild pigs, the wild goats, and other small animals. They ran away into the open country on the right side and left side, that is, toward the east and the west.

Then as the thousand men came still nearer the jungle from the south side, they began to stretch out in a long line to the right and to the [118] left. And then the men bent forward the two ends of the line in a curve toward the jungle. In that way they began to enclose the jungle, as fishermen enclose fish in a net. The men now made a still louder noise by firing their guns. At this the bigger and more obstinate animals in the jungle began to run away.

By this time the men had enclosed the jungle on three sides—the south, the east, and the west—until only the north side of the jungle was still open. And that was where the ravine started from the jungle northward.

The big animals ran along the ravine to escape from the jungle. But they did not know that the sixty hunters on the thirty elephants were waiting for them at the end of the ravine.

So as each animal emerged at the far end of the ravine, it was shot by the hunters. At first these animals were leopards, bears, wolves, and a few small tigers.

Then something wonderful happened, as I shall now tell you. In that jungle there was a big tiger and a tigress. They had recently been married, that is, the tigress had chosen the tiger as her husband—for in the jungle it is usually the wife who chooses the husband. So [119] the tiger was very attentive to the tigress. Wherever she went, he always walked with her to protect her. He also caught the prey for her, sometimes alone and sometimes with her help.

This big tiger and tigress were in the jungle, when they heard the noise of drums and guns that the men were making. Being the most obstinate animal in the jungle, the big tiger did not want to move at all. But perhaps he thought that it would be best for his wife to go away from that jungle. So she and he went into the ravine, hoping to escape.

But they too did not know that the sixty hunters were waiting at the end of the ravine to shoot them as soon as they emerged.

So the tiger and tigress walked calmly through the ravine, and emerged into the open country at the end of it.

Now I must tell you that in a tiger hunt of this kind the guest of honor has the place of danger, which was in this case right in front of the ravine. So Prince Henry waited right there on his elephant, and the hunters on the other elephants were placed in a line on his right side and left side.

This is what happened. When the tiger [120] and tigress emerged from the ravine, they suddenly saw the line of hunters blocking their path. At the same time the hunters also saw the tiger and tigress. Now I must tell you that it is a rule that only the front man on each elephant may fire his gun at once, and the man with him must reserve his shot, in case the front man misses and the tiger comes nearer. So, as soon as they saw the tiger and tigress, the thirty front men on the thirty elephants fired their guns.

But it takes at least a second for the quickest man to aim his gun and fire; and a tiger can make up his mind to do something, and do it, in less than a second. So in that time the tiger told his wife what to do.

I do not know what language tigers use among themselves, but she understood what he meant. And she did it!

This is what she did. Like a flash of lightning she leaped toward the side. So when the hail of thirty bullets came, she was not there where the hunters had aimed. Not a single bullet hit her. And in the same instant the tiger had also leaped—but onward. Some of the bullets wounded him, but not very severely, as the hunters did not have time to aim exactly.

Tiger Charging Hunting Party

He knew that he must engage the attention of all the sixty men to give his wife enough time to escape. So, wounded as he was, he leaped again, straight onward.

Then the thirty men who had reserved their shot saw a terrible sight. They saw the tiger coming straight toward the nearest elephant—Prince Henry's elephant, right in front of the ravine. The thirty men pointed their guns at the tiger. They may have vaguely seen that the tigress was escaping; but their whole anxiety was about the terrible tiger leaping straight toward them.

All the thirty men fired at him. But as the tiger was leaping onward all the time, they could not take aim properly. So if any of the bullets wounded the tiger again, the wounds were not severe.

The tiger came to the elephant on which Prince Henry was. With a huge bound the tiger leaped upward toward the box on the elephant.

So far the elephant had stood still. Being well trained, he knew that he must not move while the men on him were firing; *they* must do the fighting. But when the tiger had apparently beaten all the men and was actually [124] leaping on him, the elephant had a new duty to do: he must swerve aside. So the elephant swerved aside just as the tiger was alighting on the box on his back.

So the tiger missed his aim; instead of landing right upon the box and killing the two men instantly, his paws only reached the elephant's head. Into the elephant's head he dug his claws, and tried to scramble up.

On the neck of the elephant the mahout had been seated. He was not a hunter, but only the man who guides the elephant. So when he saw the tiger leaping upon the elephant, the mahout just dropped off on the other side, and escaped into the bushes. The tiger could have jumped down on him and killed him; but the tiger scorned to touch so humble a prey. He wanted instead to get at the hunters, who had tried to kill him and his wife.

So the tiger dug his claws on the elephant's head, paw over paw, and tried to climb up to the elephant's back. Maddened with the pain, the elephant began to rock and sway. The two men on the box could not use their guns again, as they had to clutch the box with both hands, or else they would have been thrown to the ground—then the tiger would have fallen on [125] them and killed them in an instant. The two men could do nothing to save themselves.

The fifty-eight other hunters had now reloaded their guns. Those who were nearest pointed their guns at the tiger.

"Don't shoot!" the rajah cried out. "You might hit the two men!"

That was quite true. For now the elephant was so maddened with terror and with the pain, that he was swaying, bucking, rearing. Nobody could take correct aim at the tiger.

Span by span the tiger climbed up, nearer and nearer to the box. The two helpless men in it saw the tiger's flaming eyes a yard in front of them, and they saw the tiger's fangs crashing together as if to crunch their bones.

A minute more, and these two men must die—in sight of the fifty-eight other hunters.

Then again something wonderful happened. The men could do nothing. But not so the elephant! He could do something!

The elephant recovered from his fright. He remembered all the clever tricks he had learned in his youth in the jungle, like Salar, of whom I have told you in Book I. This elephant remembered what he too could do with his trunk. [126]

So the elephant began to curl his trunk around the tiger's neck. The tiger *felt* the end of the trunk creeping around his neck.

Then the tiger knew that in the next minute the elephant's trunk would grip him by the neck and tear him off from the elephant's head; and then the elephant would bring him to the ground and trample him to death.

The tiger did not wait for that. He had scorned the sixty men—some of whom were the best hunters of the world—but he was too wise to scorn the elephant. And the tiger knew that by this time his wife must be safe.

So the tiger dropped to the ground, ran past the rear of the elephant, and vanished into the bushes. And while he did that, not one of the hunters had time even to point a gun at him.

Once only did the hunters catch sight of the tiger again. After the tigress had escaped, she must have worked her way around to the thick bushes behind the hunters; and there she must have been waiting for her husband. A few minutes later the men caught a glimpse of the tiger and tigress, husband and wife, walking together leisurely beyond those bushes, across a short open space, toward the next jungle. There they would live in the future. [127]

And as the hunters saw that sight of the tiger and tigress walking away with stately steps beyond the reach of their guns, Prince Henry took off his hat to the tiger!

"Gentlemen, I am glad that he got away!" he said to the other hunters. "I do not think that any man in history has ever charged sixty enemies single-handed, and has gained his purpose—to save the life of one dear to him."

Then Prince Henry wiped his forehead, pretending that he had taken off his hat to do that!

And so the famous tiger hunt was over. It often happens like that, in spite of sixty hunters and a thousand other men: five minutes of thrilling excitement—and then it is all over! I must tell you that if you go to hunt a tiger, even with all that preparation, you never really know whether you are going to hunt the tiger, or the tiger is going to *hunt you*! And if you do not have elephants to help you, the chances are that the tiger will hunt you.

Men, with all their guns and other inventions, can in some cases be saved from some animals only by other animals—from tigers by elephants and buffaloes, as I have described to you.

[128]

CHAPTER X

The Lion

I shall now tell you about other felines or animals of the Cat Tribe. The *lion* looks the grandest of all such animals—I suppose just because he has a *mane*.

Most lions live in Africa. There are some lions in Arabia and Persia, which are the two countries in Asia nearest to Africa. A few lions are also found in a jungle on the west side of India. These lions in the countries of Asia are not as big as the African lion.

Then there is also a species of lion in America, though he has no mane. He is called the *puma*; but people in the Western states often call him a *cougar* or mountain lion.

The puma is found chiefly in North America; he is also found in Central and South America, but not so often.

In the United States the puma lives mostly in the mountains of the Far West. He is very fond of deer flesh; and as there are still plenty of deer in the forest reserves in the Far West, the puma has managed to survive there. But in the Middle West, where there are fewer deer, there are hardly any pumas.

[129]
[130]

Group of Lions

Puma

[131]

The puma is seldom able to attack cattle. But when pressed by hunger in the winter, he sometimes descends from the mountains to the plains below, and tries at least to steal sheep from the farms.

The puma usually avoids men, especially as the men there often carry guns. But still, when made desperate by hunger, the puma has even been known to attack a man on a lonely farm.

In size this American lion, like the lions that live in Asia, is much smaller than the African lion. The African lion is the finest specimen of a lion. So I shall describe the African lion in particular.

The African lion grows to be about three feet six inches high at the shoulders; but his big head stands up quite a foot higher, and makes him look very imposing. His body, without the tail, is about five and a half to six feet long. So the African lion is not quite so long as the Bengal tiger. Still, the lion is a splendid specimen of the Cat Tribe.

"But why is the lion a member of the Cat [132] Tribe at all?" you may ask. "The lion does not *look* like a cat. The tiger does look like a cat, though much bigger than an ordinary cat."

That is quite true. But still the lion is a true cat. Why?

The Lion has the Fangs, the Tongue, the Claws, and the Paws of a Cat

You will remember what I said on page 66: that all animals of the Cat Tribe have a special kind of fangs, tongue, claws, and paws. The lion, too, has that special kind of fangs, tongue, claws, and paws; so he is a true cat. And of course the lioness has them also; so she too is a cat.

Now I shall describe these four things as possessed by the lion—or lioness.

First, the fangs. The lion or lioness has two pairs of strong fangs—one pair in the upper jaw, pointing downward, the other pair in the lower jaw, pointing upward. The lion uses these fangs in the same way that the tiger does, to hold down or to drag his prey.

[133]
[134]

African Lion
Photograph from the American Museum of Natural History, New York

Also, in chewing his food, the lion uses his fangs in the same manner that the tiger does. [135] The lion, too, has ordinary teeth, besides the fangs. So the meat lies on the lower teeth, and the upper fangs come down on the meat and pierce it. And just like the tiger, the lion, too, needs to chew his food only a few times, as the lion also has a strong digestion.

But in one thing the lion uses his fangs in a different manner from the tiger. In killing a weak prey, such as a deer or an antelope, *the lion usually bites it with his fangs on the back of the neck*. The tiger seldom kills his prey in that manner. As you will remember, the tiger usually kills an animal by striking it with his paw; and if he uses his fangs at all to kill the prey, he seizes it by the *throat* and bites it there, not at the back of the neck.

The second catlike quality that the lion has is that his tongue is rough. He can use his tongue, as the tiger does, to scrape off small pieces of meat from a bone.

The third quality of the lion like that of other cats is that the lion's claws also are retractile: that is, the lion can draw in his claws, or thrust them out, just as he pleases.

The fourth quality the lion has like all other felines is that his paws also are padded with thick muscles underneath. So the lion, too, [136] can stalk his prey silently, or harden the muscles to strike down and stun the prey with his paw, or use the muscles like springs in leaping—as I have already described to you on pages 71-72. The lion can run with a series of leaps or bounds, like any other feline.

But there is a fifth quality which all felines have, though I did not mention it before, because a few other animals also have it. This quality is that they have *sensitive whiskers*. You have noticed the whiskers of an ordinary cat. If the cat were asleep, and you touched a hair of the whiskers, the cat would wake up at once. Why? Because each hair of the whiskers is very sensitive.

All felines have sensitive whiskers; that is, each hair can *feel* any object it touches. This is a very useful quality in a feline in going about in the jungle, especially in darkness; for then the whiskers give warning of any object close at hand, by just touching it.

But, as I said, a few other animals besides felines have sensitive whiskers.

In the same manner there is another quality which all felines have, as well as a few other animals. And that quality is to be able *to see in the dark*. [137]

But it must not be totally dark. It is a mistake to suppose that a cat can see in absolute darkness. No animal can. For a cat or any other feline to see, there must be at least a tiny bit of light—even if the light is not sufficient for a human being. The eyes of the Cat Tribe are formed in such a manner as to catch the tiniest bit of light.

That is why the lion, the tiger, and all other felines can see at night in the jungle. For there are usually a few stars visible, even when there are passing clouds. Or, if the whole sky is covered with one big cloud, then the cloud itself may reflect a little light coming from various parts of the land.

But, as I said, a few other animals besides felines are able to see in partial darkness. These other animals are also night feeders or night prowlers—such as the deer, the antelope, and the hyena.

Lastly, I ought to mention one special quality which all felines have—at least they possess it more than other animals. And that is the quality of *cleanliness*. You have noticed a cat licking itself to keep clean. A lion and all other felines do that. A lion even keeps his face clean. And as he cannot lick his own [138] face, he uses his paws to clean it—just like an ordinary cat.

How the Lion is Different from Other Cats

I have told you the many qualities which the lion has like all other animals of the Cat Tribe. But can you see in what qualities the lion is *different* from all other felines?

I shall tell you. First, the lion has a mane; that is, the male animal has; the lioness has no mane. *No other member of the Cat Tribe, male or female, has a mane.*

Also, the tail of the lion has a *tuft* of hair at the end; *no other animal of the Cat Tribe has the tuft.*

Moreover, the tail of the lion or lioness hangs straight out from the body; it is not naturally *curled*, like the tail of the ordinary cat or other feline. But of course the lion can curl his tail for a moment, if he wants to,—for instance, in order to whisk off a fly.

I shall now describe to you more fully these special qualities of the lion.

The lion's mane is composed of long, bushy hair. The hair grows all around his neck, and upon his shoulders. It begins to grow when he is three years old, and continues to grow [139] till he is about five years old. A shorter growth of hair extends to the under part of the body of those lions that live in colder regions.

You may have read in your geography that in the interior of Africa there is a table-land, a part of which is about 6,000 feet high. There it is generally cold, and especially at night. So, to protect them from the cold, the lions that live there have a much thicker mane and more hair on the under part of their bodies than the lions that live in the hot lowlands nearer the sea.

When the lion lives in forest regions where there is plenty of vegetation, his mane is usually brown in color and much darker than his tawny yellow body. Why is that? Because the vegetation has both dark and yellow patches, and so the lion looks very much like his surroundings, and finds it easier to stalk his prey without being detected.

But when the lion lives in sandy or stony regions, the color of his mane is more like that of his body, that is, yellow; so he appears to be very much like the color of the sand or stones around him.

Once a lion and a lioness were drinking the water from a little pool in the stony region. [140] Two hunters happened to approach the place from behind a large boulder. They were standing about twenty yards from the lion and lioness, and yet they could not distinguish the animals. They *heard* the lapping of the water, and that is how they knew that the animals were somewhere close to them.

As for the tuft of hair at the end of a lion's tail, nobody seems to know why the lion has that tuft. The end of the tail has a hard nail, or claw, and the tuft of hair may be meant to enclose the nail, and to prevent it from being worn out against the ground. But nobody seems to know why the nail itself is there, as the lion never uses it now. Perhaps the nail had a use many generations ago, and the lion has forgotten that use now.

The tail itself, as I have already told you, hangs down straight, and does not naturally curl. It may be so because the lion does not use his tail constantly, as other animals of the Cat Tribe do, such as the tiger and the leopard. Why? Because those other animals live in denser jungles, and so they constantly use their tail as a feeler; that is, as the animal walks through the jungle his tail *feels* the objects which it touches, just like a hand; and [141] in that way the tail gives warning of any danger coming from behind. So these felines that live in the dense jungles have got used to keeping their tail stretched out like a hand; and the tail is curled upward so as not to rub against the ground.

But as the lion usually lives amid scantier vegetation, he does not need to feel his surroundings quite so constantly; and so his tail has lost the power of curling itself upward.

Of course, the lion still uses his tail to express his love or hate, as many animals do. He can express his affection by wagging his tail, just like a dog, though he seldom has any reason to show his affection for men; a tame lion, however, has actually been known to do that. But he may very often have reason to express his anger, in fact, whenever a hunter tries to kill him. Then the lion lashes his tail in anger from side to side, before leaping at the hunter.

[142]

CHAPTER XI

The Lion's Daily Life

Now I shall tell you about the remaining habits of the lion, and how he lives every day.

Lion cubs at birth are usually twins or triplets. Sometimes four or even five cubs are born together; but then they are very difficult to rear, and one or two of them usually die. So a lioness has generally a family of two or three cubs to take care of. She brings them up in almost the same way that a tigress rears her cubs, as I have already described. The lioness feeds her cubs with her milk for about the first three months, and after that she gives them a little tender meat.

When the lion cubs are six months old, they are able to eat all kinds of meat and to follow their mother to hunt the prey. She teaches them the tricks of the jungle, just as the tigress teaches *her* cubs.

So, by the time the lion cubs are about a year old, they can kill the prey by themselves. [143] Their mother just looks on, and *criticizes* their work! That is, she tells them if they have done their work well, or if they have done it badly! How does she tell them that? In this way:

If she is satisfied with their work, she does nothing in particular; she just joins the cubs in eating the prey after they have killed it. But if she is *not* satisfied with the way in which they have caught or killed the prey, she cuffs them with her paw!

Hunters have actually observed lionesses doing that! And of course the lion cubs practice their lessons more thoroughly the next time. In the jungle, the children of animals do not need to be punished more than once or twice!

You will notice that I have said nothing about the cubs' *father*, the lion. I am sorry to say that the lion is not usually so good a father as the tiger is. You will remember that the tiger helps his wife to provide food for the children, and also to teach them the tricks of the jungle. A lion seldom does that; he usually deserts his family, and lets them take care of themselves.

A lion that does stay with his family, after the cubs are born, has usually more than one [144] wife. In that respect also the tiger is far finer than the lion. A tiger has only *one wife*; and he takes care of her and the cubs. But when a lion does stay with his family, the family usually consists of two or three lionesses, who are his wives, and their cubs.

In that case they hunt the prey in a pack; that is, the lion and the lionesses all hunt the prey together; and they are even helped by the older cubs. They need to hunt in a pack when the prey happens to be large, such as a buffalo or a giraffe. A lion by himself could seldom kill a buffalo or a giraffe.

Many a fight has been observed in the jungle between a lion and a buffalo—and almost every time the buffalo has succeeded in driving off the lion with its horns. Even if the lion managed to leap upon the buffalo from the back, he could not kill the buffalo by *biting it on the neck* because of the thick hair there.

And if the lion tried to stun the buffalo with a blow of his paw on the buffalo's head, the blow would not be enough, because of the thick hair which grows on the African buffalo's head. And meanwhile the buffalo would rear and buck, and throw off the lion. But if the lion has one or two lionesses to help him, they can [145] all attack the buffalo at the same time, and pull it down and stun it with many blows.

On the other hand, as you will remember, in a fight between a single tiger and a single buffalo, the tiger always wins; he dodges the buffalo's horns, then seizes the buffalo by the *throat* from underneath. In that way he always kills the buffalo. It is only a herd of buffaloes that can beat a tiger, not just one buffalo.

A lion by himself is also unable to kill a giraffe in most cases; for if the giraffe sees the lion coming, it will kick out with its hind legs or its fore legs; and a kick from a giraffe has been known to disable a lion completely. So if a lion by himself wants to attack a giraffe, he must first stalk the giraffe stealthily, and then jump on it suddenly.

But as the lion cannot usually come near enough to do that, he generally attacks a giraffe with the help of one or two lionesses. For then they can all attack the giraffe from different sides; and as the giraffe cannot kick different ways at once, one of them is sure to jump upon the giraffe's back and bite it on the neck.

As I have just said, a lion cannot often [146] stalk his prey near enough to leap upon it. There is a reason for that. Compared with his size, *the lion's leap is the shortest of all members of the Cat Tribe*. The farthest that a lion has been known to leap, even with a run, is about thirty feet—whereas a tiger has been seen to leap a distance of forty-eight feet!

The lion's body is not meant for leaping far. His chest and fore legs are very strong, but his hind legs are not quite so strong—and in leaping an animal uses its hind legs most. For instance, the kangaroo has the biggest leap of all four-legged animals of its size; and it has very large hind legs and very small fore legs.

"But if the lion cannot leap very far, how does he catch his prey at all?" you may ask.

I shall tell you. Like all other felines, he usually hunts at night. He hides near a pool or a stream, and waits for his prey to come to drink. Then he tries to approach the prey noiselessly on his padded feet. If he succeeds in creeping near enough to leap upon it, he certainly has his meal that night. But if he does not succeed in doing that, he tries another plan. He roars!

[147]
[148]

Giraffes

Kangaroo

And that is an advantage a lion has over all [149] other animals. None of them can roar like him. Even a tiger's roar is not so loud, and so he seldom tries to roar. But very often a lion *must* roar to catch his prey, and so by constant practice he has made his roar very terrible indeed.

Yes, the lion really catches his prey by roaring. When the animals are drinking at the pool, the lion puts his mouth to the ground and roars. It sounds just like thunder.

When you hear a roll of thunder, it sometimes happens that you cannot tell from which direction the thunder is coming. In the same way, when the animals hear the lion's roar, they cannot always tell from which side the roar is coming, because by putting his mouth to the ground the lion sends the roar in all directions. So in their terror some of the animals run the wrong way, and actually run toward the lion. Then the lion finds it easy to leap upon at least one of them.

The lion seldom hunts in the daytime. But when he does, he uses a different method. He chooses a pool amid sandy or stony ground. Then he half buries himself in the sand, or lies low among the stones and boulders. So if any animal comes to drink from the pool, it does [150] not notice the lion because the lion's tawny color makes him look like the sand or stones. Then the lion leaps upon the animal and catches it.

After having his meal, the lion drinks from the pool. If the prey is rather large, so that he cannot finish it at one meal, he keeps it for the next day's meal. He drags the animal's body to some hiding place and covers it up with sand or leaves. Of course, he stays somewhere near that place, as otherwise *the thieves of the jungle* would eat up the food. The thieves of the jungle are the jackal and the hyena.

But as the lion usually hunts his prey in the night, he generally sleeps in the daytime. He is not really dangerous except at night. If a man meets a lion suddenly in the daytime, the lion will not usually attack him, unless very hungry. Many a man who has met a lion in the jungle by day has escaped in safety by just standing still, making no sound and no motion. After a glance at the man, the lion has walked off.

Most wild animals are afraid of man. Perhaps that is because they do not quite understand him, or how he can hurt them from a distance [151] —by shooting them with a gun or even with an arrow. That is why most wild animals try to avoid man, unless they are wounded or are very hungry.

But I must tell you here that a tiger attacks a man much more readily than a lion does. Even in the daytime a tiger will usually attack any man he meets—like the fisherman that the tiger carried off from the river, as told on page 110.

At night, however, *all* animals of the Cat Tribe are dangerous, and many a night a lion has been known to creep into an encampment and carry off a sleeping man. That is, the lion first killed the man, then *dragged* him away.

In that respect a lion is different from a tiger. A lion usually takes away his prey by *dragging* it; he grips his victim in his jaws by an arm, or by the shoulder, or by the neck, so that the victim trails along the ground.

A lion once seized a sleeping man by the wrist, and dragged him away. The lion thought that he had killed the man. But the man was still alive. He got up on his feet as he was being dragged away. He *walked* by the side of the lion for a few yards; meanwhile he drew his revolver from his pocket [152] with the other hand, and then shot the lion through the head, killing him instantly.

A lion seldom carries his prey *bodily* as a cat carries a mouse. A tiger always does that, if the prey is light, like a man; and a heavier prey he actually carries over his shoulder—as I have said on page 103.

From all the facts I have told you so far, you will understand that a tiger is stronger than a lion. It has been reckoned that the strength of a lion is equal to that of five men, but a tiger's strength is equal to that of eight men. How that was calculated I shall tell you in another book.

A tiger is also much more ferocious and terrible an animal than a lion. The lion can be hunted on horseback; the tiger must never be hunted in this way. A hunter riding a horse has often come to within a hundred yards of a lion, and has killed the lion with one or two shots from his gun—and the horse has stood quite still while he took aim.

But a horse will never face a tiger or stand still before a tiger. The horse will be in a panic at the very sight of a tiger—and will flee in terror. Even if a band of horsemen meet a [153] tiger, all the horses will stampede in terror. It needs an elephant—a trained elephant—to face a tiger, as I have already described to you. And usually it needs several elephants to *hunt* a tiger.

The tiger has also many more of the catlike qualities than the lion has. The tiger is more active than the lion, can leap farther, and can make up his mind more quickly. Above all, like a cat, the tiger has "nine lives." Many a time a hunter has killed a lion with a single shot. But usually it needs half a dozen shots even to disable a tiger.

If a lion is mortally wounded through the heart or through the head, he usually drops to the ground at once. But if a tiger were mortally wounded in the same manner, he would at least leap toward the hunter, and try to kill his slayer, before he himself agreed to drop down and die.

The lion has sometimes been called the King of the Jungle—I suppose because in those countries where he lives there are no tigers. So the lion is the "monarch of all he surveys" in his own jungle. Of course, the lion looks grander and more imposing because he has a mane, and the tiger has none. Perhaps that [154] is the reason why some people have given the lion that title.

The lion has also been called a noble animal, but accounts differ as to his real character. Sometimes a lion has behaved very splendidly, as in the two stories I shall tell you presently. But, on the other hand, there have been occasions when a lion has behaved like a coward and a sneak, as people have declared. So I suppose that lions are like other creatures: there are good lions, and there are bad lions.

In one respect, however, the lion is much finer than the tiger: the lion can be tamed, but the tiger cannot. At least, we can say for certain that many a lion has been known to become quite tame, but never a tiger.

There was an actual case where a tiger was caught as a small cub and brought up on milk, and then on clean meat without any blood on it. The tiger grew up, and was thought to be quite tame. Then one day, as he was licking his master's hand, his rough tongue drew blood from the hand—and in a moment, at the sight of the blood, the tiger became a ferocious wild animal.

Luckily, a faithful servant crept from behind with a gun, and suddenly shot the tiger [155] through the head. The master leaped out of the room at once, before the tiger could reach him in his dying struggles.

But as for the lion, not only can he be tamed, but even a wild lion has been known to behave as if quite tame, when moved by his love. I shall now tell you two stories about that.

[156]

CHAPTER XII

The Lion a Noble Animal

Androcles and the Lion

Many, many years ago, the Romans ruled a large part of the world; for they were a great nation. Their territories included the north of Africa. A rich Roman, who lived there, had many slaves. One of his slaves was called Androcles (An´ drō clēz). The Roman treated Androcles very cruelly. So Androcles ran away from him.

But the Roman sent out many soldiers to capture Androcles. So after hiding in many places, Androcles was at last compelled to flee into wild regions, where there were few inhabitants. As the soldiers followed him even there, he had to go still farther into the interior of the country, till he came to the jungle. There he lived by eating fruits.

One day, toward evening, he was sitting on the ground, when suddenly he saw a lion before him. Poor Androcles gave himself up for lost, [157] as he had no weapon with him with which even to try to fight the lion. He knew it was useless to try to run away, as the lion could catch him with a couple of bounds. So he thought that his only chance was to sit quite still, for then the lion *might* go away.

But the lion looked at him, and then came toward him. The animal did not rush toward him or leap. Instead, the lion just walked toward Androcles.

That was strange, Androcles thought. The lion came nearer and nearer—and then Androcles noticed that the lion walked in a peculiar manner. That puzzled Androcles. But he sat quite still, hoping that the lion would yet go away.

But instead the lion came right up to him. *Now* he would be eaten up, poor Androcles thought.

Then a wonderful thing happened. Instead of eating him, the lion held out a paw toward him. Then Androcles understood.

He looked at the lion's paw closely. He saw that the paw was swollen. Yes, that is why the lion had been *limping*.

Androcles took the paw in his hands and examined it. On the under side he found a [158] large thorn embedded deep in the flesh. It must have been there for several days, and must have caused the lion intense pain.

Androcles pulled out the thorn carefully; then he squeezed down the swelling. That relieved the lion's pain.

Immediately the lion showed his gratitude. He wagged his tail, fawned on Androcles, and gambolled around him playfully like a dog. He could not do more to show his feelings.

After a time the lion went away to the jungle.

A year passed. Androcles still lived in hiding. Then at last he was captured by the soldiers, and brought before the judge.

It used to be the law in those days to condemn runaway slaves to death. Also, it used to be the custom to put to death Christians and condemned slaves by casting them to lions.

So one afternoon all the Romans in that place were gathered to make a holiday. It was a kind of circus they had come to see, only, instead of having the usual clever tricks which you now see in a circus, the Romans had fights between men and men, between men and animals—and finally, as a grand finish, the Christians and the condemned slaves were thrown to wild lions. Many of the lions had [159] recently been captured from the jungle; so they were quite wild. And as they had been kept without food for two or three days on purpose, they were very ferocious and quite eager to eat the Christians and the condemned slaves.

When it came Androcles' turn to be eaten, he was thrown into the enclosure, which was called an arena. Then a wild lion, which had been recently caught from the jungle, was let loose into the arena from a cage.

Ten thousand Romans looked on to see Androcles die. And Androcles looked up to the Romans, and found no mercy in them. He looked at the famished and furious lion—and knew that he must die.

For the lion crouched ten yards before him, lashing his tail in fury. The lion gave a bound, and came within five yards of Androcles.

There the lion crouched again for a moment—then made a rush at Androcles. Everyone thought that *now* the lion would kill Androcles.

But a still more wonderful thing happened. Instead of killing Androcles, the lion gambolled around him, and fawned on him—as if he were glad to meet again an old friend.

Then Androcles understood. He had for [160] gotten all about the lion he had met in the jungle the year before, whose pain he had relieved. But the lion had not forgotten *him*.

Who says that animals have no memory? This lion had a memory! He carried in his memory the gratitude of his heart for the pain that Androcles had relieved. Although Androcles was now dressed differently—in fact, most of his clothes had been stripped from him—the moment the lion had drawn near enough to him, he had recognized Androcles as his old friend and benefactor of the jungle.

Famished as he was, and furious at being kept without food, the lion would gladly suffer the pangs of hunger rather than injure a hair of his friend's head. Instead, the lion fawned on him, then lay down before him like a lamb.

Then something melted in the cruel Romans' hearts; perhaps they realized that there was some Great Power beyond them, who had inspired a raging beast of the jungle to be as gentle as a lamb.

The Romans asked Androcles to explain this marvel. He told the story of his adventure with that lion in the jungle—just as I have told it to you.

[161]
[162]

Androcles and the Lion
 [163]

Then Androcles was pardoned, and given his freedom, in memory of this great wonder.

My dear children, this story has a special meaning for us. We are told that if we cast our bread upon the waters, it shall be returned to us. That means that if we do an act of kindness, we shall have our reward. Androcles did an act of kindness to the lion in the jungle. In return Androcles was given back his life in the arena.

The Lady and the Lioness

I shall close this chapter by telling you another true story. It happened quite recently, in America. In a zoo there was a lioness. She had two little cubs. She was very fond of them, and she used to lick them with her tongue many times every day to keep them clean. They used to trot around her and scramble over her, then lie down beside her, one on each side, to have another cleaning with her tongue.

One day the lioness and her two cubs were lying like that quite close to the bars of the cage. One of the visitors there happened to be a man who had an umbrella. Very foolishly he poked one of the cubs with the umbrella. [164] He did not mean to hurt the cub; I suppose he only wanted to *feel* it. But still it was very foolish to poke the cub with the umbrella.

In an instant the lioness jumped up with an angry roar, and thrust out her paw between the bars. Luckily for the man, she could not quite reach his arm; otherwise she would have dragged him to the bars of the cage and killed him instantly. Instead, she could only reach the umbrella. So she seized the umbrella, and wreaked her vengeance on it. She smashed it to a thousand bits. The man, of course, ran away.

Then gradually the lioness quieted down. She lay down as before in front of the bars, with the cubs beside her, one on each side. Now and again she gave them an affectionate lick with her tongue, first one, then the other. That helped to sooth her feelings somewhat. Still, as you may well understand, she was bitter at heart at the foolishness of some people.

Now it so happened that a lady had observed the whole incident. She had been standing all the time in front of the cage, a few yards away. And this lady had two little girls with her, one four years old, and the other six years old. [165]

You may be sure that the lioness saw the lady and the two little children. After a time the lady came a little nearer to the cage, the two little girls standing beside her, one on each side. The lady tried to catch the lioness's eye. Presently their eyes met. While the lioness was still looking at her, the lady patted her two little girls on the cheek.

Then the lady came a step nearer the cage. As the lioness licked her cubs, the lady patted her own little children; and she smoothed their cheeks and hair.

The lioness saw that.

The lady was just waiting for that. She came still nearer to the cage. Each time the lioness licked her cubs, the lady stroked the cheeks of her own children affectionately.

Then the lady began to speak. She spoke in a very soft voice, very gently and very slowly. She spoke softly as if she meant only the lioness to hear her. This is what she said:

"I at least understand you. I too am a mother, like you. See, these are *my* two children! I love them as you love yours."

Then the lady took up the children, one on each arm. She kissed the children, first one, and then the other—and the kiss seemed [166] almost like the act of the lioness in licking the faces of her own cubs. By that the lady meant the lioness to understand that the children were just the same to her as the cubs were to the lioness.

Then the lady spoke again, as softly and tenderly as before:

"My children also love your children. Wouldn't it be nice if they could play together!"

Then the lady held the smaller girl in front of her. Very timidly the little girl held out her hand—while her mother looked into the lioness's eyes.

Well, my dear children, I cannot tell how it happened. Perhaps some message of love and sympathy and understanding passed between the two mothers—the mother of the two little girls, and the mother of the two little cubs. At any rate, this is what actually happened:

Very timidly and very slowly the lady stepped to the cage. The little girl put her hand between the bars, and petted the cub nearest to her. The lady moved a little, and the girl petted the other cub. The lioness looked on all the time.

Then something still more wonderful hap [167] pened. As the little girl was petting the cub, the lioness also began to lick the cub; then the lioness's tongue passed over the cub's body and came to the child's hand—and *the lioness began to lick the child's hand as if the child were her own.*

Remember that this was a wild lioness, and untamed. Nobody had ever dared before even to come within her reach.

Then the lady turned a little, and brought the other girl to the bars of the cage—and she too petted the cubs. Lastly, the lady put the girls down, and passed her own hand through the bars. She too petted the cubs, then finally she stroked the lioness herself.

And that was like a kind of handshake as a good-bye. They parted friends—like two mothers who had met by chance on the roadside, and each had admired the children of the other.

[168]

CHAPTER XIII

The Leopard

The *leopard* is another animal of the Cat Tribe. You may know him at once by the *spots* on his body; and of course the female leopard also has the spots. These spots are usually black in color, or sometimes very dark brown. But the color of the body, or "ground color" as it is called, is different among the several kinds of leopards.

For, I must tell you, the leopard lives in so many countries that he varies in size and in ground color in different countries. He is found in almost all parts of Africa. In Asia he lives mostly in the hot countries in the south; but a special kind of leopard, called the snow leopard, is found in the cold countries in the north of Asia. On the American continent there is also a kind of leopard, called the jaguar.

Now I shall describe in detail all the qualities of these different kinds of leopards. [169]

First, as leopards are felines, they have the fangs, the tongue, the claws, and the paws of the Cat Tribe, which I have already described to you.

The Leopard's Ground Color and Spots

The ground color of the leopard's skin is usually yellow, but the shade of yellow varies in different leopards; sometimes it is a bright yellow, sometimes a brownish yellow. There are leopards whose skin is even darker than that,—some actually black.

"But why do different kinds of leopards have different ground colors?" you may ask.

Because they live on different kinds of soil and amidst different kinds of vegetation. You will remember what I have already told you: that *the color of an animal's body is very often the same as the color of the place where he lives*. Then the animal's prey or enemy is not able to distinguish him from his surroundings. So the animal finds it easier to catch the prey, or to escape from the enemy. And, because the color of the soil and vegetation in different countries varies from yellow to brown, the color of the leopard's body also varies in that manner, at least as a rule. [170]

Now I shall tell you about his spots, which are always of a dark color. But they vary in shape in different kinds of leopards. In some leopards the spot is a solid round disc, like the shape of a coin.

In other leopards the spot is like a thick ring; that is, there is a gap at the center. In some leopards the ring is broken up in parts; that is, the ring is not a complete line, but is made up of a number of short lines. The spot then looks like a rosette, because these lines spread outward like rose petals.

All these that I have just mentioned are regular shapes. But in many leopards the spots are quite irregular.

The spots also vary in *size*. In some leopards the spots are larger than a silver dollar, and in some they are as small as a quarter-dollar.

Why the Leopard has Spots

Now you may wonder why different kinds of leopards have different kinds of spots, both in shape and in size. I shall tell you. Each has the kind of spot that is most useful to him. How is that? How can the spots on the leopard's skin be *useful* to him? Why does the leopard have spots at all?

[171]
[172]

Leopard

Jaguar
Photographs from the American Museum of Natural History, N. Y.
[173]
First, I must mention that all leopards can climb trees, just like cats.

People believe that once upon a time lions and tigers could also climb trees. Of course, they climbed only big trees, which have a very thick bark into which they could dig their claws deep enough to bear their weight. But now the lion and the tiger have forgotten how to climb trees. Perhaps they did not keep up the use of their power to climb trees.

But the leopard has kept up his habit of climbing trees. In fact that is the way *he usually catches his prey*. Does not that seem wonderful? I shall explain how he catches his prey in that way.

He chooses a tree near a stream, or near a pool of water, where different animals come to drink. The leopard climbs up to a bough of the tree, about ten or twelve feet from the ground. He lies flat on the bough and waits.

Presently a deer comes to the water to drink. The leopard waits till the deer is quite near, perhaps actually passing under the bough. Then suddenly the leopard jumps down on the deer and catches it.

The leopard often does that in the daytime, as well as at night. And in the daytime [174] the sun may be shining, and on some nights the moon may be shining. It is *then* that the spots are useful to the leopard. Can you tell why?

Because when the sun or the moon is shining, a little of the light peeps down between the leaves of the tree and reaches the ground. Have you ever noticed that? If so, you have seen that the light reaches the ground like little *bright* spots, but that there are little *dark* spots also — the bright spots being the little patches of light peeping down, and the dark spots being the shadows where the light is shut off by the leaves.

In the same way there are bright patches and dark patches on the bough of the tree, where the light also falls in that manner.

And that is what a leopard's body looks like — bright patches and dark patches. The dark patches are his spots, and the bright patches are the ground color of his skin.

So if the deer did happen to look up to the bough when approaching the tree, it would not be able to distinguish the leopard from the natural patches of light and shadow near by. So the deer would not notice the leopard, and would be caught. [175]

And that is why the leopard finds his spots so useful to him in catching his prey.

But why do different kinds of leopards have different kinds of spots? Because they live in different countries, which have different kinds of trees; and so the patches of brightness and darkness made by the sunlight or moonlight are also different.

[176]

CHAPTER XIV

The Leopard's Habits

Now I shall tell you the other qualities and habits of the leopard.

First, his *size*. The leopard is smaller than the tiger; he is not quite three feet high at the shoulders. The length of the leopard's body, without the tail, is about five feet.

That is the average size of the *male* leopard. In describing each kind of animal I am usually telling you about the male, because he is generally larger and stronger than the female. Why? Because the male has to do the fighting to protect the family, especially when the children are very young.

The leopard's *strength* is so great that he can break a steer's neck with a blow of his paw. He cannot carry a steer on his back, which a tiger can do, but still the leopard can drag the steer for some distance. As for a deer, the leopard can easily carry it. That has been discovered in a strange manner. As I have [177] told you, a leopard lies on the bough of a tree and waits for a deer to pass under the tree. One time a leopard happened to kill a deer in that way. As he was not very hungry, he ate only a few mouthfuls from the throat and from the under part of the deer.

He wanted to keep the deer for his next big meal. But if he kept it on the ground, the jackals and hyenas would find it in his absence and eat it up. So what did the leopard do? Can you guess?

Well, the leopard carried the deer up that tree, and placed it crosswise on the fork of the bough. Then he climbed down and went for a prowl. He knew that the thieves of the jungle—the jackals and the hyenas—could not climb the tree and steal his dinner.

But a party of hunters passed that way and saw the deer's body on the fork of the tree; and they knew that a leopard had carried it up there.

How could they know that? Very easily. The hunters brought down the deer's body and examined it. They found that the deer's throat and under part had been eaten.

Now I must tell you that hunters know from the study of the jungle that each wild animal [178] has a different way of eating its prey. A leopard always eats first the throat and the under part; but a tiger always eats a hind leg first. So these hunters knew that it must be a leopard that had eaten the deer's throat and under part.

And the hunters also knew before, from their study of the jungle, that a leopard can climb trees; but they knew that more certainly after this incident. How? Because they knew from the deer's throat that a leopard had killed it and partly eaten it; and they *found the deer in the tree*. So they concluded that the leopard must have climbed the tree and hidden the deer there.

This also proves the fact that the leopard is really an intelligent animal. The lion and the tiger hide their prey by merely placing it in a hollow in the ground, and covering it loosely with sand or leaves. But unless the lion and the tiger are very watchful, the thieves of the jungle often steal their dinner; that is, the jackals and the hyenas smell the flesh, and uncover it and eat it up.

But the leopard hides his prey more securely. As he has the power of climbing trees, he uses that power to carry his prey to the fork of a [179] tree, where the thieves of the jungle cannot reach it.

My dear children, there are many people who do not use the natural gifts they have. The leopard does better than that. He uses his gift of climbing trees in two ways: first to *catch* an animal passing beneath, and then to *hide* the prey in the tree. Had the lion and the tiger continued to use their former gift of climbing trees, they too would have been able to hide their dinner safe from the thieves. Instead, they now find it stolen many a time, and have to go hungry.

The leopard, of course, uses his other gifts in catching his prey in various ways. Being a feline, he too can give a big bound like a cat, and as he also has padded feet, he can catch his prey by stalking it. He creeps silently through the jungle, till he comes near his prey; then he gives a sudden bound and falls upon it.

The leopard has splendid muscles; the muscles are not big, but they are hard. The leopard leads such an active life that he is generally slim, without any flabbiness. In fact, the leopard is a perfect type of feline grace, beauty, and agility. The lion is the laziest animal of the Cat Tribe; the leopard is the [180] most active. The leopard is even more active than the tiger.

The Panther: Popular Name for Large Leopard

There is no such animal as the *panther*. That is only the popular name for a large leopard—particularly a large and ferocious leopard.

Some people fear a large leopard even more than they do a tiger, because a large leopard attacks a man even more often than a tiger does. Other wild animals as a rule avoid man, as I have told you before. But a tiger very often attacks man, and a large leopard does so almost every time he can. He is by nature even more ferocious than a tiger.

The leopard has this very bad quality: he is perhaps the only animal that kills for the mere "fun" of killing—just like some men who call themselves "sportsmen." If a large leopard gets among a herd of cattle, he kills several of them, one after another. He does the same with wild pigs, wild goats, and wild sheep. He kills many more than he can possibly eat. In fact, the bad name some people give to the tiger in that respect really belongs to the [181] panther or large leopard. When a large number of animals are found killed, a tiger is usually blamed for it.

But wise people, who have studied the ways of animals, never make that mistake. Of course, they cannot always tell by the paw marks on the ground whether a small tiger or a large leopard did the killing—because the paw marks of a large leopard look so much like those of a small tiger. But if a single one of the animals killed has been eaten, then they know whether it was a tiger or a leopard that did the killing. How do they know that? By examining the part eaten—as I have already described to you on page 178.

How the Leopard Seizes his Prey

A leopard usually seizes his prey by the throat. He grips the throat in his jaws, and holds on till the animal cannot breathe and is suffocated.

If the prey is large, such as a big stag, the leopard's grip on the throat may not suffocate it completely; then the leopard uses another method. He keeps his grip on the throat of the prey, and *pulls downward* with his full weight. The prey tries to rear up on its hind legs to throw off the leopard—but then the leopard pulls downward with a sudden jerk. This breaks either the animal's spine or its neck, and it falls to the ground.

The leopard seizes his prey by the throat when it is a swift-footed animal, like the deer. But when it is a slow-footed animal, like cattle, the leopard uses another method—at least on some occasions. He rushes to the prey from the side or the back, and kills it by a blow of his paw on the neck from above—as a tiger does. If one blow only stuns the prey, and it falls, the leopard just starts *eating* the throat, which of course kills the prey.

The Leopard's One Amiable Quality—He Loves Perfumes

The leopard is said to have at least one amiable quality. It is said that he is so fond of beautiful perfumes that he can be tamed with them! That is, if you use some beautiful perfume which the leopard likes, you can tame him with it for a time. But I cannot tell you whether that is *always* true.

There are many things said about animals that are not always true, for instance, that every animal can be charmed with music—if only we use the particular kind of music which that particular animal likes. No doubt, particular kinds of animal *have* been charmed in that way for thousands of years; and even the most terrible kind of snake, called the cobra, is regularly charmed in India with a flute.

You must have read of these serpent-charmers in storybooks, as they charm even *wild* cobras in that way. So it is quite true that several kinds of animals can be charmed with particular kinds of *soft* music, such as the music of the flute and the violin. I shall tell you all about that in my next book.

But about taming leopards with perfumes—we are not sure that *all* wild leopards can be tamed with beautiful perfumes. It is at least true that *some* wild leopards have been tamed in that way. I shall now tell you a true story, to show you how that once happened.

The Leopard and the Lavender

Once a wild leopard had been caught in a trap in the jungle. He was put into a cage and carried overland to a seaport. There the leopard in his cage was put on a ship to be taken to England. The cage was placed on the deck of the ship. [184]

The leopard was very wild and ferocious. If any of the passengers or crew came anywhere near the cage, he snarled with rage and leaped at the bars of the cage. He shook and bit the iron bars, as if he wanted to get out and attack the people. He was well fed all the time, but still nothing seemed to lessen his ferocity.

Then, one day, a lady happened to take out her handkerchief. She was standing about three or four yards from the cage, and a fresh breeze was blowing from her direction toward the cage. Immediately a change came over the leopard. A minute before he had been snarling with rage at sight of her, and trying to get out to attack her.

But as soon as she took out her handkerchief, the leopard ceased to snarl and to bite the bars. Instead, he tried to put his head through the bars, as if to get his *nose* as near her as possible.

Of course the lady did not understand that. She merely wondered why the leopard had changed his behavior so suddenly. She now noticed that the leopard was bending down, and scratching the floor of the cage near the front of the bars—just as a pet cat or dog will scratch the floor outside your door [185] to be let in. The lady wondered still more, and came a little nearer to the cage.

Immediately the leopard got up, and began pacing the cage in joy. The lady now stood about two yards away. Then the leopard put his paw through the bars and began to *snatch* with it. The lady was a little frightened at first, but presently she noticed that the leopard was not snatching at *her*, but at the *handkerchief*, which was still in her hand. And the leopard was not snatching ferociously, but almost playfully, like a great big cat.

After a moment's thought the lady realized that the leopard wanted the handkerchief—but why he wanted it, she did not know. So she threw the handkerchief at the bars. The leopard caught it in his paw, and pulled it into the cage.

Then you should have seen how that wild and ferocious leopard behaved! He played with that handkerchief more joyously than any kitten ever played with a ball. He put the handkerchief on the floor of the cage, leaped upon it, rubbed his nose on it, and even rolled over it.

Gradually the lady began to understand why he did that. The handkerchief had been scented [186] with lavender. She wondered if it could be the *lavender* that he loved, and not the handkerchief itself?

Struck by this idea, the lady went to her cabin and brought out a small bottle of lavender scent. She opened the stopper, and splashed a few drops of the scent through the bars. Then the leopard simply went crazy with delight. He leaped upon the places on the floor where the drops had fallen, and he rubbed his nose on them, and rolled over them. Then the lady knew that it was the scent that the leopard loved.

After that she gave him the lavender to smell every day, and the leopard became so tame that he allowed her to come to the bars and pat his body.

But as this is a true story, I must tell you the ending. One of the men passengers on that ship gave the leopard a large piece of cotton-wool soaked in lavender. That was unfortunate — I mean it was unfortunate that the man used cotton-wool instead of a handkerchief or even a piece of cloth.

The leopard played with the cotton-wool in delight, and rubbed his nose and face on it. In doing so he must have got the cotton-wool [187] into his mouth — and then he must have taken in a deep breath. We don't know whether he meant to do that, as he liked the perfume so much, or whether he took the breath without meaning to do so. In any case, the cotton-wool got into his windpipe, and he tried to cough it out; but he could not. The foolish passenger did not know what was the matter; and so he did nothing.

Then in a few days an inflammation set in, and the poor leopard died. Some people are so thoughtless!

[188]

CHAPTER XV

American Leopard: The Jaguar

Now I shall tell you about an American leopard. He is called the *jaguar*. He lives mostly in Central America and South America. His favorite country is Brazil, near the Amazon and other rivers that flow into the Amazon.

Some people call the jaguar the American *tiger*. This is a mistake, because a tiger is striped, not spotted; and the jaguar is spotted, like a leopard. So it is more correct to call the jaguar the American *leopard*.

He has all the qualities of other leopards that I have already described to you. But his spots are a little larger and not quite so completely round; they are more nearly square, with rounded corners.

All four-footed animals can swim naturally in some fashion, but leopards can swim especially well. And the jaguar, who lives near the Amazon and other rivers, is a champion swimmer. He swims as easily as he climbs [189] trees. So he eats fish as often as he eats monkeys!

Yes, he actually catches a monkey sleeping on the bough of a tree! He climbs up so silently that the monkey does not awake. At least, those monkeys that do not cultivate the keenest sense of hearing, even in their sleep, get eaten by the jaguar. But a jaguar that is clumsy in his movements awakes the sleeping monkey—and then that jaguar has to go without his dinner. So, again, life is like a competition or trial in the jungle, as I have told you in Book I, pages 118-119. Those animals that cultivate their gifts escape their enemies and they get enough to eat. Those that do not cultivate their gifts are either killed by their enemies, or are themselves starved to death.

The jaguar is very fond of monkey for his dinner, just as you are fond of roast turkey. The things he likes next best are fish and turtle. He catches a fish by pouncing on it from the bank. Turtles that he finds on the bank he merely turns over on their backs, so that they cannot run away. Then he leisurely scoops out the flesh with his paws and eats it.

But when the jaguar is in the water pouncing [190] on fish, he in turn has an enemy that wants to eat *him*. When the jaguar has pounced on a fish, a silent snout may come up to him from behind—and grab him! Yes, an alligator! And the alligator needs only to hold the jaguar in his jaws, and drag him down, and keep him under water till the jaguar is drowned. Then the alligator can have jaguar flesh for *his* dinner.

Here again we have an example of competition in the jungle. The jaguar must cultivate not only quickness in catching fish, but also his own sense of hearing, so as to escape from the alligator in time.

"But what about the alligator?" you may ask. "Doesn't *he* need to cultivate some gift to escape his enemy? Is there no enemy that tries to eat the alligator in his turn?"

No! There is no other animal in the water that wants to eat the alligator, or that can do so. But still the alligator may have an enemy near by, who wants to kill him. There may be a hunter on the bank who wants to shoot the alligator to provide you with purses, handbags, or satchels. So the alligator too must be on his guard against his own enemy.

[191]
[192]

The Chain of Conflict in the Jungle

You can understand the whole story by supposing that there are in that place: [193]

A small fish,
A bigger fish,
A jaguar,
An alligator, and
A hunter.

Then let us suppose that the small fish is trying to catch some tiny creature of the water on which it feeds. But while the small fish is catching the tiny creature, the fish itself must look out for its own danger. Otherwise:

A bigger fish comes, and eats the small fish. But the bigger fish also must look out for its own danger. Otherwise:

The jaguar comes, and eats the bigger fish. But the jaguar also must look out for his own danger. Otherwise:

The alligator comes, and eats the jaguar. But the alligator also must look out for his own danger. Otherwise:

The hunter comes, and shoots the alligator.

So you see that the animals that dwell in the jungle have to cultivate all their gifts to get on in life.

[194]

CHAPTER XVI

The Dog Tribe

I have told you of several flesh-eating animals that are of the Cat Tribe. But there are some flesh-eating animals that are of the Dog Tribe. The most important one of these in the jungle is the *wolf*.

How can you tell the difference between the Cat Tribe and the Dog Tribe? By the four qualities that the Cat Tribe has, and which the Dog Tribe does not have.

I. The members of the Cat Tribe have four fangs. Those of the Dog Tribe do not have fangs. They have special teeth of their own kind.

II. The members of the Cat Tribe have a rough tongue. Those of the Dog Tribe have a tongue which is not quite so rough. They do not need a very rough tongue, as they can scrape the meat from a bone with their teeth.

III. The members of the Cat Tribe have retractile claws. The Dog Tribe's claws are [195] rigid and stiff; that is, they are thrust out all the time. The members of the Dog Tribe do not use their claws in seizing or holding their prey; they hold the prey in their jaws.

IV. All of the Cat Tribe have padded paws: they have them for many reasons, which I have mentioned on pages 71-72. But the paws of the Dog Tribe are not so thickly padded with muscles. The Dog Tribe do not need the thick padding of muscles, because:

1. They do not need to stalk their prey silently. They catch their prey by running it down, as a greyhound catches a hare.

2. They do not strike down their prey with their paws, but seize it in their jaws.

3. They do not need to give a *bound* in catching their prey, so the muscles under their feet need not act like *springs*.

The members of the Dog Tribe gain on their prey by moving their legs *quickly*, not by covering a large amount of ground with each movement of their legs. But the Cat Tribe do just the opposite: they do not move their legs so quickly, but they cover a larger amount of ground at each movement of their legs. As I have told you already, a dog *gallops*, but a cat *bounds*. The dog's legs move much faster [196] than the cat's, but the cat gives a bigger jump than the dog each time.

The American Gray Wolf

I have said that the most important wild animal of the Dog Tribe is the wolf. Wolves are found in every continent—Europe, America, Asia, and Africa. And there are many species of wolves in these continents. I shall tell you more about them in another book, but now I must tell you about the American gray wolf.

There is in the United States one of the most wonderful animals in the world—the American gray wolf. He is perhaps the only animal in the world *that has beaten man*!

I mean this: Man has killed off many four-footed wild animals; that is, he has killed so many of those animals in a place, that they have *died out* in that place. He has not succeeded in killing off the American gray wolf.

In some places man has almost killed off certain animals, even when he did not *want* to do so. He killed the animal merely for sport or for profit—but he did not want that species of animal to die out altogether; for then he could not have any more sport or profit from it. And yet, the hunter killed so many of that species of animal that it has almost died out in some places. In this manner, as I have already told you, almost all the elephants have been killed off in parts of Africa, for the sake of sport or for the sake of the tusks. In the same way, the buffalo has almost disappeared from the United States.

[197]
[198]

Gray Wolf
From a photograph copyrighted by the New York Zoological Society.

[199]

But in the case of the American wolf, man *wanted* to kill him off altogether as a race of animals; and yet he has not been able to do so. At first the hunter may have killed the wolf only for the sake of its fur; but in the last few years the American farmer and the ranchman have tried to wipe out the wolf altogether as a *pest*— because the wolf kills their sheep and cattle. And yet, the wolf flourishes in the West. He has beaten the farmer and the ranchman.

The wonderful part of it is that the American wolf has beaten man *by his own efforts*. And for an animal to beat man in that manner is a great achievement.

I have told you before that one animal has to use its gifts against another animal, to protect itself from danger; for instance, the fish has to look out for the jaguar, and the jaguar in his turn has to look out for the alli [200] gator. But in that competition of the jungle, the animal has generally to use its wits merely against another animal—not against man. But the American wolf had to use his wits against man; and he has beaten man, as I shall describe to you.

The American Wolf Learns to Evade the Gun

About a hundred years ago, when people began to go West, they shot many buffaloes, wolves, antelopes, and deer. They did that for sport or for profit; they made a profit, because they sold the skins and other parts of the animals' bodies. At that time the hunters did not want the animals to be killed off altogether, but they actually killed so many of these animals in a few years that the buffalo, the deer, and the antelope became scarce. These particular animals, of course, tried to use their wits to escape from the hunters. They did not succeed in doing so. They fell as victims of the gun.

But not so the wolf. He began to use his wits against man and his gun. He soon realized that man was his enemy and also that man could kill him from a considerable dis [201] tance. A wolf saw a man at a distance; then the wolf heard a bang, and immediately felt a sharp pain in his body. That wolf fell and died. But another wolf saw his brother die like that. He set his wits to work. He concluded that the man had caused the bang which made his brother fall and die. Hence the wolf realized that man was dangerous to him, even at a distance. So after that the wolf resolved to run away from man. And other wolves learned to do the same.

Of course, the whole race of wolves did not learn this lesson so quickly. Many hundreds of wolves meanwhile fell victims to man's gun; but a few wolves escaped. These few wolves also saw repeatedly that if any of their brothers allowed a man to approach anywhere near him, he was killed. So after seeing that happen many times, the surviving wolves learned that they must always run away from the presence of man.

These few surviving wolves taught their children to do the same. Some of these wolf children did not heed that lesson when they grew up; so they too were killed. But a few of the wolf children remembered the lesson when they grew up; so they escaped getting killed. [202]

In turn these wolves also taught *their* children to run away from the presence of man. So in a few generations a race of wolves grew up in the West that the hunter *did not even get the chance to shoot.*

That in itself was a great achievement for the wolf. Why? Because some species of animals as a race do not learn so quickly to run away from the mere presence of man; one or two animals personally may learn quickly to do that, but not all the animals of a species. That is why the buffaloes and some of the antelopes and deer in the West were wiped out; they did not learn in those same few years to run away from the presence of man. The wolves alone learned this, and they have survived as a race.

The American Wolf Learns to Evade the Trap

But the battle was not yet over. Seeing that his gun had now failed, man used his wits to kill the wolf in another way. He set *traps* for the wolf; and he cunningly baited the traps with tempting food. Then the man went away from the traps. He thought that because he was not himself anywhere near the [203] traps, the wolf would not be afraid to approach them. Well, at first some wolves did go up to the traps, and were caught by them.

But a few other wolves saw that fate of their unwary brothers. So those surviving wolves again set their wits to work to discover the cause of this new danger. And after a time they saw the steel traps. "So, *this* is our new enemy!" they said.

After that they avoided the traps, even if the traps were baited with the most tempting food. And they taught their children to do the same.

So again man was beaten in this battle of wits. He found that the trap could catch the wolf no more.

But man tried again. He *hid* the trap cunningly under leaves or under snow; only the tempting bait was placed in sight. He thought that because the wolf could not now *see* the trap, he would fall into it.

Well, some wolves did fall into it.

But a few other wolves saw the fate of their unwary brothers. So these surviving wolves again set their wits to work to discover a way of detecting the traps. Perhaps they saw the hunter's footprints; or perhaps they [204] realized that the snow or the leaves covering the trap did not look *natural*. You remember, in Book I, how Salar's father detected a very tricky trap because the ground there did not look natural. Well, in some way, the surviving American wolves detected the traps, even when the traps were covered up. So after that they began to avoid these *hidden* traps, and they taught their children to do the same.

Man found himself beaten once more by the wolf in this battle of wits. He found that the American wolf could not be caught even by a hidden trap.

That again was a great achievement for the American wolf. Why? Because even the elephant, clever as he is, gets caught at last by a tricky trap, even if he avoids it for a long time. To do better than the elephant is a triumph indeed!

So far the hunter had tried to kill the wolf for the sake of the fur; and the wolf took no revenge for these years of persecution. He bore no grudge against man, and did not try to pay him off. The wolf merely wanted to live, and to be let alone. Man would not let him alone. He wanted to kill the wolf just for the sake of money. [205]

Then a new thing happened. Many people began to go West; farms and ranches began to be started. These farms and ranches had many sheep and cattle.

Then the wolf had *his* turn! He found that sheep and cattle were far easier to kill than the wild animals on which he had made a living so far. So the wolf began to raid farms and ranches at night. He still avoided man; he never let a man come near enough to shoot him; and he never touched a hidden trap. But still he began to kill sheep and cattle.

Man now found the tables turned on him! Formerly he had persecuted the wolf; now the wolf persecuted, or at least tormented, *him*! So man made one last desperate effort to beat the wolf in this battle of wits.

The American Wolf Learns to Evade the Poison

Man set his wits to work, and at last devised the use of *poison*. He selected different kinds of poison, with different tastes and different smells, — or no taste and no smell at all! He chose the nicest kinds of meat, on which to put the poison. Then he cunningly placed pieces of the poisoned meat all over the paths by [206] which the wolves must come to raid the sheep and cattle. He thought that *now* he would beat the wolf!

Well, some of the wolves did eat the poisoned meat; they died. But a few of the wolves saw the fate of their unwary brothers. So these surviving wolves once more set their wits to work to discover the cause of this new danger. It may have taken them some time to suspect that the meat was the cause of this new danger; and a few more wolves may have died meanwhile from eating the meat.

But some of the wolves did detect the new danger. We do not know exactly how they did so. Perhaps this time they used one of their other gifts to save their lives; that is, they used their power of *smell*. They recognized man's scent in or about the meat. So they knew that man had put the meat there.

They had long known that anything that had to do with man was dangerous to wolves. So the wolves resolved to leave the meat untouched. Instead, they went on raiding the sheep and the cattle. And they taught their children, and their children's children, to do the same.

And now the American wolf has beaten man, [207] finally and absolutely. The farmer and the ranchman can think of no other method of killing the wolf. So the American wolf continues to flourish merrily.

The marvel of all this is that the wolf is not naturally a very intelligent animal. Most animals have far more natural intelligence than the American wolf; and yet none of these animals seem to be able to beat man in the battle of wits. The American wolf alone has done it, though he naturally has very little brains.

But *he has used all his brains*. He has concentrated his efforts to save his life by beating man. He has not only used all his brains, but he has done so *all the time*. He determined to overcome each new danger as it arose. And he *worked hard all the time*.

My dear children, that is a great lesson for us. All children, or all men and women, do not have great talents; but everybody can use all the brains he or she has. Some few people prosper in life because they have talents and use them. Other people of talent are lazy, and do not use all their gifts; these people do *not* prosper. But many people, who have no talent at all, prosper just the same; they do what the American wolf has done. [208]

1. They first decide on something *worth doing,* just as the wolf decided on saving his life.

2. Then they *use all the brains they have* to do that thing.

3. They *concentrate* their efforts on it.

4. They *work hard all the time* to do that thing.

5. As they meet each difficulty or danger or trap, they devise a method of *overcoming* that difficulty or danger or trap.

If you learn this much from the American wolf, you will learn the secret of success in the battle of life, when you grow up.

Meanwhile, remember all that I have told you, till I come back and tell you in the next book many more Wonders of the Jungle.

Till then, as they say in the Orient, God and His peace be with you!